METAL/POLYMER COMPOSITES

METAL/POLYMER COMPOSITES

John Delmonte

Delsen Testing Laboratories
Glendale, California

Van Nostrand Reinhold
New York

Copyright © 1990 by Van Nostrand Reinhold
Library of Congress Catalog Card Number 89-30815
ISBN 0-442-22100-2

All rights reserved. No part of this work covered by the copyright hereon may be reproduced or used in any form or by any means—graphic, electronic, or mechanical, including photocopying, recording, taping, or information storage and retrieval systems—without written permission of the publisher.

Printed in the United States of America

Van Nostrand Reinhold
115 Fifth Avenue
New York, New York 10003

Van Nostrand Reinhold International Company Limited
11 New Fetter Lane
London EC4P 4EE, England

Van Nostrand Reinhold
480 La Trobe Street
Melbourne, Victoria 3000, Australia

Nelson Canada
1120 Birchmount Road
Scarborough, Ontario M1K 5G4, Canada

16 15 14 13 12 11 10 9 8 7 6 5 4 3 2 1

Library of Congress Cataloging-in-Publication Data
Delmonte, John
  Metal-polymer composites / John Delmonte.
    p. cm.
  Includes index.
  ISBN 0-442-22100-2
  1. Metallic composites. 2. Polymeric composites. I. Title.
TA481.D45 1989
620.1'18--dc19   89-30815
                                                                CIP

*To Those Individuals Whose Vision
Extends Beyond the Horizon*

PREFACE

The growth of composites is most evident when comparing materials of the early 1960s with materials currently being developed, and their applications. In a book on metal-filled plastics, which I published in 1961, the focus was upon combinations of finely divided metals with synthetic resins. Although such combinations continue to be employed, the available materials, both metals and polymers, have not only increased dramatically, but the techniques of effecting combinations have become more sophisticated. Vacuum deposition, ion implantation and chemical deposition, for example, have supplemented mechanical blending. As a consequence, consumer electronics in the guise of magnetic tapes and semiconductive materials is opening new horizons for metal/polymer composites, in areas which were little known twenty-five years ago. Continuing research in thin film technology and semiconductors are in some instances concerned with molecular electronics, where the behavior of materials no longer follows patterns characteristic of macrostructures. Photopolymers and their applications as photoresist coatings on semiconductors aremaking possible new developments in the nanostructures of microelectronic devices.

Large scale applications of metal/polymer composites are now commerically viable in the automotive, aerospace and aircraft industries. Advanced composites used in these fields include not only new metallic alloys, but also fibrous reinforcements of glass, silicon-carbide, graphite, polyaramids, aluminum oxide, silicon carbide, fine metal fibers and metal whiskers, metal-coated polymer fibers and flakes. New developments reflect the manufacturing ingenuity of the United States and Japan as well as the contribution from Great Britain and West Germany. The author has decided to recognize some of these activities by drawing attention to recent patent developments and recent technical papers which underscore the wavefront of continuing technology.

The expanding role of metal powders, which are adapting some plastics manufacturing techniques to the fabrication of powder metal compacts is covered in the early chapters. New amorphous metal compounds formed by rapid cooling solidification processes, extend the rapport between metals and polymers.

Many individuals have contributed suggestions and ideas which appear in this book and, to these individuals, I express my gratitude. In particular, I wish to acknowledge Dr. Jack Ching and Lee McCrory, president of Delsen Testing Laboratory, both of whom reviewed some chapters, and to Professor Robert Grubbs of the California Institute of Technology who reviewed Chapter 4 on electroconductive metal/polymer composites. I am also grateful for helpful suggestions by Raymond Seymour, Distinguished Professor of Polymer Science, particularly for Chapter 5 on coatings. I have drawn upon my diverse experience on materials over a period of fifty years, and any mistakes that may have been introduced are my personal responsibility. The patience and help of Bonnie Sidewell and Jacquie Thibault in the typing and retyping of a complicated manuscript are gratefully acknowledged. To my wife, Janet, my loving thanks for making possible the required environment for a writing effort.

It is my hope that this book will stimulate interest in composites by exposing the plethora of opportunities that lie ahead.

John Delmonte
Glendale, California

CONTENTS

Preface / ix

1. **Introduction / 1**
   Historical Perspectives of Metal/Polymer Composites (MPC) / 2
   Metal Wire Reinforcement / 5
   Metal Fibers/Polymer Fibers / 11
   References / 11

2. **Production of Finely Divided Metals and Polymers / 12**
   Production of Metal Powders / 15
   Characteristics of Metal Powders / 17
   Manufacturing Processes for Metal Powders / 19
   Production of Powders from Polymers / 26
   Combining Finely Divided Metals and Polymers / 28
   References / 38

3. **Molding and Casting of Metal/Polymer Composites / 40**
   Pressure Molding of Plastics Particles / 41
   Pressure Molding of Metal Particles / 44
   Molding of Combined Metal and Polymer Particles / 49
   Similarities of Processing Metal and Plastic Powders / 52
   Basic Contributions of Metals and Plastics
     to Molded Composites / 52
   Encapsulation of Metal Components by Plastic Materials / 54
   Metal Inserts / 58
   Metal Powders and Liquid Polymers / 60
   Miscellaneous Metal Pastes / 69
   Interactions between Metal Particles and Polymer Particles / 70
   References / 76

x   METAL/POLYMER COMPOSITES

4. **Electroconductive Polymer/Metal Composites / 77**
   Trends in Electroactive Polymers Development / 79
   Synthesis of Electroactive Polymer Development / 81
   Polyacetylene and Polythiophene Developments / 84
   Deposition of Metallic Elements by Thermal Decomposition / 85
   Metallic Ion Implantation / 86
   Photodielectric Analysis / 87
   Electrical Conductivity in Metal Filled Composites / 89
   Carbon Blacks / 97
   References / 99

5. **Plastics Coated Metals and Metal Coated Plastics / 102**
   Reinforced Coatings on Metals / 104
   Non-Solvent Mastic Coatings / 105
   Plastisols / 108
   Dispersion Coatings / 111
   Plastic Coating of Metal Structures / 113
   Powder Coatings / 115
   Sprayed Metal Coatings on Plastics / 117
   Electroplating Metals on Plastics Surfaces / 119
   Polymer Plating on Conductive (Metal) Substrates / 121
   Vacuum Metallizing on Plastics / 123
   Ablative and Intumescent Coatings on Metals / 132
   References / 133

6. **Metal/Polymer Structural Composites / 135**
   Filament Wound Structures / 146
   Pultrusion / 148
   Polymer Impregnation for Structural Purposes / 153
   Low Density Metal Structures / 155
   Metal Matrices / 159
   Metal/Plastics Combinations / 160
   References / 161

7. **Radiation Shielding by Metal/Polymer Composites / 163**
   EMI (Electromagnetic Interferences) and
     RFI (Radio Frequency Interferences) / 163
   Comparison of Technologies for Shielding / 174
   Control of Static Discharge / 178
   Plastics Packaging for Food Products / 181

Coaxial Cable Shielding / 183
Shielding against X-Rays and Nuclear Reactors / 184
References / 184

8. **Metal/Polymer Composites in Magnetic Components / 187**
   Magnetic Materials / 189
   Developments in Permanent Magnets / 193
   Magnetic Material Development for Transformers, Motors and Generators / 196
   Magnetic Materials in Communication / 197
   Miscellaneous Magnetic Materials of Polymer/Metal / 205
   Magnet Wire Enamels / 206
   References / 208

9. **Micro and Nano Electronic Applications / 210**
   Semiconductors / 211
   Integrated Circuits / 222
   Packaging for Electronics / 223
   Storage Batteries / 228
   Capacitors / 232
   Photovoltaic Cells / 234
   References / 237

Glossary / 239
   Terms Used in Metal Powder and Metal Processing / 239
   Terms Relating to Plastics/Polymers / 240

Index / 243

METAL/POLYMER COMPOSITES

1
INTRODUCTION

In an earlier book, *Metal Filled Plastics*, published in 1961, I discussed the consequences of adding finely divided metals to plastic polymers to create qualities not usually found in the plastic products. The roles of organic polymers in the impregnating and bonding of porous assemblies of metal powders or fibers were also examined. Control of density, increase of electrical conductivity, and decorative effects were obvious characteristics which were emphasized. Commercial products using these properties multiplied and are still being created today.

Basic research in electroconductive polymers and semiconductive metals have narrowed the differences between metals and organic polymers allowing many more combinations of materials to be formed. Synergism of material combinations is now the focus of attention, yielding advanced composites with unique properties. These are the materials preparing us for the twenty-first century. So vast is this proliferation of materials that it is appropriate to define the limitations of what will be pursued in this book.

Developments in rapidly solidified metals and non-metals (with the rapid cooling of molten metals in a fraction of a second) yield amorphous non-crystalline substances which possess unique physical and electrical properties. The literature reveals hundreds of "new" material combinations some of which show remarkable strengths when processed further into films or fibers. Amorphous metal powders combined with substantially non-crystalline organic plastics will also allow the formation of "new" materials.

Although the subject of advanced composites definitely includes ceramic materials (in addition to metals and organic polymers) emphasis here is placed upon the inter-relationship of metals and polymers. In current research on organic polymers, there is a strong focus on establishing those characteristics which make polymers more electroconductive and more heat resistant. On the other hand, the metal powder associations are identifying the advantageous processing techniques of the plastics industry using metal powders, particularly in precise molded and sintered parts, and the concomitant benefits of improved physical and electrical properties.

In assembling the data in this book, the following guidelines were used:

(1) emphasis upon metal/polymer combinations and their processing; (2) concentration upon molecular aggregates rather than the unique chemical entities of the many thousands of metallo-organic compounds; and (3) select applications of metal/polymer composites to emphasize trends in technology.

New relationships between metals and polymers are being established at microscopic levels. These relationships and interactions will become apparent as semiconductors, integrated circuits, and photovoltaic devices are examined in later chapters. The new technologies of thin-film semiconductors; ion-implantation, to place metal-ions into selected ceramic and polymer substrates; vacuum metallization of plastic films; magnetic particles in polymer films in magnetic tape recordings; amorphous metals, as produced by rapid solidification; and adaptation of plastics processing techniques, are but a few of the burgeoning activities establishing roles for metal/polymer composites. These technologies provide more than the addition of a metal component to a plastic assembly or vice versa. They create an intimate relationship of two or more materials to attain a superior performance, previously unknown in the exploitation of the earth's raw materials. A brief review of the historical perspectives follows.

HISTORICAL PERSPECTIVES OF METAL/POLYMER COMPOSITES (MPC)

The proliferation of composites that has occurred during the past one hundred years may be gleaned from a review of composites and material origins over the past several thousand years.[1,2] The expression "composite" first appeared early in the twentieth century, and is now generally used to describe the union of two or more diverse materials to attain synergistic or superior qualities to those exhibited by the individual members.

Early cultures found it advantageous to use two materials to attain a superior product such as a weapon or tool. Thus, for example, the bone or ivory handle on a metal dagger provided a more easily handled weapon. Thin films of hammered gold leaf decorations, for wood or leather products, yielded more attractive articles. Wood, leather, and bone are examples of early composites, as they are derivatives of natural polymers. Cellulose, a polysaccharide of glucose, is present in vegetable matter in substantial quantities. Collagen is the major protein component of connective tissue in mammals. As the aging process occurs, it becomes molecularly cross-linked. By tanning collagen, leather products are formed.

Specific examples of early historical metal/polymer composites of gold, silver, copper, and iron are dated from the periods when those metals became available. Copper artifacts have been found in the Near East dating from the third millennium B.C., and exhibit animal forms shaped over wood models.

Metal Coated Polymers. Metal coated polymers have attained wide-spread recognition today in the preparation of packaging products, electronic devices and decorative films. From the past, there is only one material that survives—the thin films of gold leaf which have adorned the leather on the covers of centuries old books, and as ornamentation on royal palaces and places of worship. Burmese and Thai temples (Fig. 1-1) are typical examples of the ancient practice of enriching and enhancing the durability of these venerable structures with gold films. Many centuries ago metal craftsmen discovered the ease with which pure gold could be hammered into fine films, which could be placed over works of art, wood carvings, and metal idols.

Wheels of Chariots. Historians usually acknowledge the development of the wheel to have occurred four to five thousand years ago. In the ceramic cultures of Mediterranean countries and in China, the advantages of rotating the clay during the formation of pottery were carried over into rotating wheels for transportation vehicles. Rotating wheels for carts and chariots, adapted the potter's wheel to broader horizons. With the appearance of metals, it became apparent that excessive wear on the rotating surfaces of the hub could be coun-

Figure 1-1. Gold leaf encrusted temples have existed in Burma and Thailand for hundreds of years.

After the development of bronze alloys in the second millennium B.C., bronze artifacts were in wide use in the Near East and in China because of their strength and processing advantages. Iron and steel production and artifacts of these materials may be traced to the early smelting of iron ore in ancient Turkey (Armenia and Anatolia).

An example of an early copper artifact from one of the temples of Ur (Ancient Mesopotamia—ca. 2000 B.C.) was reported recently.[3] A copper statuette inscribed in Sumerian cuneiform was enclosed within a mud brick box and sealed with bitumen pitch (a natural occurring polymer from oil deposits of southern Mesopotamia). The inscription was deciphered by Dr. N. Kramer of the University of Pennsylvania, establishing the early time period. The copper artifact had added significance because it was embedded in a corner foundation stone of a major Sumerian temple. This discovery provides one of the earliest *dated* combinations of metals and polymers. Other references to historic metal/polymer composites (MPC) were summarized by Delmonte.[4]

The following examples of metal/polymer composites are of historical significance. They are representative of unique combinations of metals and polymers used to achieve improved quality of the products. Before the twentieth century, most organic polymers were of natural origin and when exposed to adverse environments, such as moist climates and ultraviolet radiation, their tendencies to degrade over long periods of time contributed to their deterioration. Wood, as a living tree, has remarkable longevity over many centuries. Animals and plant matter trapped in the frozen waters of arctic regions have also maintained their integrity. Few natural polymer/metal composite examples have survived. Exceptions are those composites maintained in a dry desert climate (as in Egypt) or by exclusion from sunlight and oxygen (when buried in a moist bog, as in Denmark).

Polymer Coated Metals. To preserve appearance and enhance performance, it is quite probable that early artifacts of bronze and steel received thin coatings of polymers. Natural resin polymers from plants were available, including rosin, shellac, tung oil, seal blubber, pressed oils from nuts and seeds, bees wax, paraffins, bitumen pitch and asphalt, fossilized copal resins, ox blood, gummy extracts from papyrus, lac, extracts from boiled residues of fauna. These represent a few of the potential film forming resins or polymer precursors available from ancient records. As noted above, most of these were subject to biodegradation and hence have not survived over the centuries. In a later section of this book, newer synthetic polymers such as polymethyl methacrylate and polyvinyl fluoride are noted for their durability to weather exposure. While today's technology and chemistry have contributed much to enhance the quality of thin polymer films, they too, will not necessarily endure for thousands of years.

teracted by the insertion of metal bearing surfaces such as bronze. Hence, in museums today, one can find examples of bronze bearings on the hub of the wheels. These were particularly evident in the horse drawn chariots of Roman warriors. Wood structures, comprised of cellulosic polymers and metals at the rotating pivot points are examples of successful historical metal/polymer composites.

Weapons and Tools. Early developments in weapons and tools received much impetus from bronze alloys and steel products which made available tough, strong, and sharp cutting edges. Metal tools and weapons, though badly corroded in some instances, have survived. Many of these used a non-metallic handle to provide better "feel" and dexterity in their use. Shaped bone pieces, carved ivory tusks, and carved hard woods fulfilled personal needs and were readily attached to metal inserts. For ceremonial purposes, applied decorations of gemstones, precious metal inserts, or colored mineral pigmentation added value and quality to the tool or weapon. Jewel bedecked daggers and swords are referred to in literature.

Armor and shields for protection of warriors in the Middle Ages had undergone extensive development in Western European countries. They were unwieldy and cumbersome and the need for mobility in battle raised questions about their usefulness, although they were designed to thwart the thrusts of spears and swords. Originally armor consisted of heavy leather or quilted fabrics, but as the fabrication skills of steel shapes evolved, metal armor began to appear during the eleventh and twelfth centuries A.D. Interlaced chain mail and conical helmets were used in the Battle of Hastings. Descriptions of the lighter weight steel mail coats have appeared recently (Ref. 1-5). Leather and fabric layers were inserted to provide flexibility and comfort for the wearer. Forging of all the steel components and the application of identification markings contributed to a complex metal/natural polymer composite garment with articulated joints. Figure 1-2 shows a composite suit of armor from the second half of fifteenth century A.D.

METAL WIRE REINFORCEMENT

The making of wires is one of the ancient metal-working crafts. Early recognition of the utility of a drawn-out bar or rod of metal for the purposes of ornaments, tools, hooks, or reinforcements prompted the manufacture of wire. It is quite likely that malleable native gold, silver, or copper lent themselves to being hammered and drawn out. Copper alloys, including bronze and brass, as well as lead alloys have been used by metal craftsmen for over 2000 years. The working of steel products required higher temperatures and pressures, and wire products of iron and steel are generally thought to have orginated in the

6 METAL/POLYMER COMPOSITES

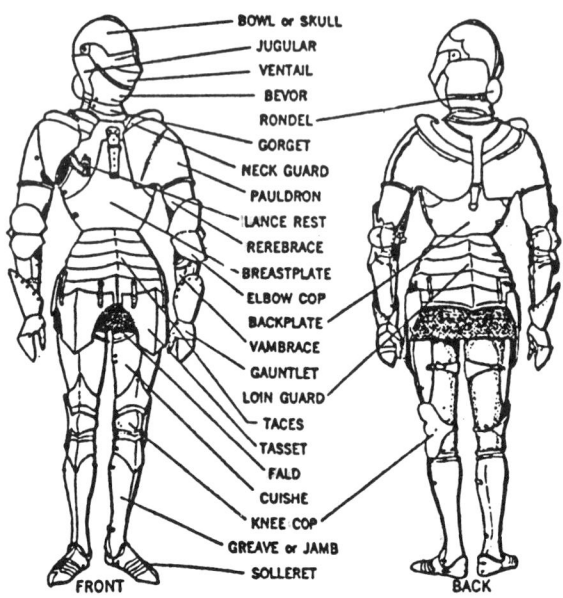

Figure 1-2. Composite metal armor of the fifteenth century. Courtesy of the Metropolitan Museum of Art.

fourteenth century. Wire drawn at Nuremberg was formed by heat treatment and the pulling of the metal through lubricated tapered draw plates. Refinements in improving alloys, coatings, and techniques to furnish metal wires followed in the succeeding years. This background is cited briefly to suggest that metal wire decorations and reinforcements were available in the Middle Ages and that combinations with natural resins and coatings were inevitable.

Up to the beginning of the twentieth century, barrels of primitive guns were reinforced by circumferential wrapping of steel wire about the cylinder for added strength. If these methods were applied to a wooden gun barrel a metal reinforced polymer would result. The use of fine silver wires as decorations on wooden gunstocks, a feature of some weapons in France and Germany during the eighteenth century, is also to be noted. In the twentieth century, the reinforcement of rubber tires with metal wires contributed to longer lasting wheels. Gutta-percha, natural rubber, benefited from these combinations. Further examples will be cited as pultrusion and extrusion processes are noted in a later chapter.

Fine wire reinforcements in the manufacture of art objects known as cloisonné have been used for many centuries. Copper and gold wires, soldered to a metal base, were shaped into exquisite designs, filled with colored enamels, and baked. While many of these products utilized wire reinforcements with colored clay slips, these practices have been supplemented by organic matrices in recent years. (Figs. 1-3A, B).

The see-through advantages of a transparent polymer matrix about shaped wires have added to the appeal of recent cloisonné, as it continues to reflect the metal working skills of Chinese craftsmen. Filigrees of fine silver and gold wires have enhanced the exquisite jewelry and handicrafts of Eastern and Western Mediterranean countries, though unlike the cloisonné mentioned above, the wires were not encased in polymers or ceramics.

Early Uses of Powdered Metals. Metal working craftsmen undoubtedly collected small quantities of powdered metals from the attrition, machining, and grinding of metal bases, and uses for powdered metals, particularly in combination with polymer matrices have appeared within the last 100 years. A page from an early U.S. patent is reproduced in Figure 1-4. It describes the mixing of finely divided metals such as lead, aluminum, tin, zinc, etc., in vulcanized rubber compounds and the use of lubricants. After the newer synthetic resins first appeared in the 1930s, combinations with metal powders proliferated. These are subjects pursued in more detail in later chapters. In passing, it should be noted that the extraction of aluminum from bauxite became a commercial process only in 1886, whereas the availability of copper, iron, lead, zinc, and tin may be traced back more than 2,000 years. Noble metals such as gold, as noted earlier, were probably among the first to attract craftsmen. Powdered gold has been used as pigmentations for inks and paints both in Europe and in China.

Metal Inlays in Wood Furniture. During the sixteenth and seventeenth centuries A.D., particularly in Italy and France, metal inserts were both mechanically fastened and/or bonded to the surfaces of wood paneling for decorative purposes. The artistic results were attractive (Fig. 1-5) and are highly prized as museum displays of furniture. The artisans displayed a sense of creativity in preparing table tops with assorted wood grains forming decorative wood mosaics, sometimes in combinations with metals and colored lithics. Because of the sensitivity of wood structures to changes in moisture (and hence changes in dimensions), problems arose in ensuring the ability of the metal inserts to remain in place. Periodic maintenance measures were adopted to establish moisture barriers through the use of waxes, oils, and resinous coatings. In addition to metal inlays on wood, there have been metal inlays and imprints upon leather products and book bindings which have stood the test of centuries of use. These rare books are maintained in major libraries throughout the world.

8 METAL/POLYMER COMPOSITES

(A)

(B)

Figure 1-3. Wire reinforced cloissone composites have been used in China for centuries.

UNITED STATES PATENT OFFICE.

HARRY A. HOFFMAN AND WALTER H. JUVE, OF AKRON, OHIO, ASSIGNORS TO THE B. F.
GOODRICH COMPANY, OF NEW YORK, N. Y., A CORPORATION OF NEW YORK.

ART OF MIXING RUBBER WITH COMMINUTED METAL.

1,395,413.   Specification of Letters Patent.   Patented Nov. 1, 1921.

No Drawing.   Application filed March 31, 1920. Serial No. 370,256.

To all whom it may concern:

Be it known that we, HARRY A. HOFFMAN and WALTER H. JUVE, citizens of the United States, residing at Akron, in the county of Summit and State of Ohio, have invented a certain new and useful Art of Mixing Rubber with Comminuted Metal, of which the following is a specification.

This invention relates to the art of mixing finely-divided metals, such as lead, aluminum, tin, zinc, etc. into plastic substances such as raw rubber compounds. In vulcanization, the petrolatum is absorbed in the rubber, probably forming what may be termed a solid solution of the two materials, and the metal particles are isolated in the final product by the vulcanized rubber which may contain, around each particle, a substantial quantity of the petrolatum, the degree of isolation depending upon the relative amount of metal present in the mixture, and the ratio of petrolatum to rubber being of course greatest immediately adjacent to the metal particles. Considerable difficulty has heretofore been experienced from the tendency of the metal particles to mass together and unite in lumps or sheets of the metal when mixed with the rubber by a dry milling operation on a roller mill or other form of rubber masticator. The rubber, when vulcanized, has been found to be weakened, and uniformity of its physical properties prevented by the unequal distribution and coalescing of the metal particles. In the manufacture of golf balls, for example, where comminuted metallic lead is mixed with one or more of the rubber elements of the ball to increase the weight of the latter, it has been found that the lead particles have united in flakes, weakening the rubber and impairing its flexibility, the stiff flakes of metal in some instances breaking out through the surface of the sheet of rubber when the latter is formed in place in building up the ball.

Our invention has for its objects to preserve the original degree of fineness of the metal particles, prevent their massing together in cakes, flakes or sheets in the milling operation and secure their more uniform distribution throughout the mass of rubber.

We accomplish these objects by coating the individual metal particles with a thin film of a lubricant which will cling to the particles and keep them separated during the mixing operation. A suitable coating lubricant is petrolatum. The coating may be accomplished by dissolving a relatively small amount of petrolatum or the like in a volatile solvent, such as benzol or gasolene, and treating the finely-divided metal with this solution. The solvent is then removed by evaporation. When the solvent has been removed, the petrolatum remains in the form of a thin film on the individual metal particles and keeps them separated throughout the mixing operation.

We have obtained good results by treating 104 pounds of 350 mesh lead with a solution comprising 2¾ pounds of petrolatum dissolved in a half-gallon of benzol. This mixture is thoroughly stirred and the excess benzol evaporated in a shallow vessel. The final trace of benzol is removed with a vacuum drier, and the material is then ready for compounding with the rubber by the usual milling methods.

We do not wholly limit ourselves to the materials and mode of procedure above set forth.

We claim:

1. The method of physically combining comminuted metal with gum plastics which comprises coating the metal particles with a lubricant and milling them into the plastic.

2. The method of mixing finely-divided metal with rubber which comprises coating the metal particles with a thin film of petrolatum, and milling them into the rubber.

3. The method of mixing rubber with comminuted lead which comprises coating the lead particles with an oleaginous material, and working them into the rubber by mastication.

4. The method of preparing comminuted metal for mixing with a plastic material which comprises mixing the particles with a lubricant dissolved in a solvent, and evaporating the solvent.

5. The method of mixing rubber with comminuted lead which comprises mixing the lead particles with a solution of petrolatum, evaporating the solvent, and milling the coated particles into the rubber.

6. A composition of matter comprising finely-divided lead coated with petrolatum.

7. A composition of matter comprising vulcanized rubber through which is distrib-

Figure 1-4. An early U.S. patent on metal powders.

10   METAL/POLYMER COMPOSITES

Figure 1-5. Metal inlays in wood—Italy, sixteenth century. Courtesy Museo Poldi Pezzoli, Milan.

A silver jacketed bible hand inscribed in the sixth century A.D., may be seen at Upsalla University in Sweden.

Underwater archaeologists have recovered artifacts of early metals and wood from the ocean seabeds. Under such non-oxidative environments, combinations of wood, leather, and metals have been found in ancient shipwrecks off Cyprus. The copper artifacts were badly corroded. On the other hand, large bronze bells recovered from tombs near Wuhan (Heibei Province—China) still fastened to decomposed wooden support posts are over 2000 years old. They have been restored to operation, as the writer was privileged to observe.

Of more recent significance is the raising of the Swedish warship, Wasar (sunk in 1628 A.D.). This vessel, recovered in recent years, is now on display in a museum in Stockholm. It contains fine examples of 300 to 400 year old composites of metals, wood, pitch, fibers of hemp, etc. When buried in ocean or lake bottoms, and deprived of oxygen, remarkable preservation of age old artifacts has occurred. The most obvious metal "finds" on the Swedish vessel were cannons and anchors.

METAL FIBERS/POLYMER FIBERS

Fabrics woven from natural fibers and from finely drawn metal wires (gold in particular) have appeared among the handicrafts of Asian countries. Fabrics woven of colored silk with gold threads have adorned the garments of wealthy people for hundreds of years and are representative of special uses of metal/polymer composites. The practice of preparing unique, expensive fabrics continues to the present, and suggests that the weaving techniques of the textile industries are receptive to fine fibers of any source, metals, ceramics and polymers. In today's commercial activities, metallic fibers may be included with organic fibers for the purpose of negating electrostatic charges. This development has appeared on carpets. It is unlikely, however, that old fabrics took account of electrostatic charges. The presence of metal threads was purely decorative.

REFERENCES

1. Delmonte, J., SPI Composites Institute, Section 3D, Cincinatti, (February 1987).
2. Delmonte, J., *Origins of Materials and Processes.* Lancaster, PA: Technomic Publishing, (1985).
3. Anon., *Expedition* Philadelphia, PA: University of Pennsylvania, (July 1987).
4. Delmonte, J. "History of metal polymer composites," Anaheim, California: ACS, (September 1986).
5. Peirce, I. "Armes et armoures," *Dossier Historie et Archaeologic,* No. 117, p. 42. France, (July 1987).

2

PRODUCTION OF FINELY DIVIDED METALS AND POLYMERS

Metal powders and polymers are of prime importance in the preparation of metal/polymer composites. Techniques for metal powder preparation have increased dramatically in recent years and will be reviewed in this chapter. Particularly noteworthy are materials produced by rapid solidification processes (RSP). There are different techniques used in preparing finely divided polymer particles. Metal powders may be blended with polymer particles in a variety of ways, though thermal blending in extrusion equipment and blending into liquid polymer systems are also practiced. The material blending equipments are also reviewed in this chapter.

Reference will be made to standard screen size and comparisons on the basis of micrometers (μm). For sub-micro dimensions, nanometers will be used as well as angstroms. Thus, dimensions of fine powders will refer to micrometers (1×10^{-6} meters); nanometers (1×10^{-9} meters); and angstroms (1×10^{-10} meters). A quick reference that will be useful is one mil (.001 inch) which equals 25 micrometers. To keep these small particles in proper perspective, one must remember that atomic dimensions are of the order of several angstroms. Fine powders (usually one micrometer and larger) comprising metal/polymer composites for molding entail colonies of metal atoms or molecules and colonies of polymer molecules. Below one micrometer in size, as metallic particles approach one nanometer, the property changes which occur in the final composite are not always anticipated. This is particularly apparent in composite magnetic devices discussed in a later chapter. When the subject of continuous thin metallic coatings arises in later chapters, sub-micron thicknesses will frequently be encountered and surface oxidative reactions and particle migration phenomena will be more prevalent.

Fine metallic particles under one micrometer in size appear in many new United States and Japanese patent disclosures for applications in composites found on electronic, electromagnetic, and semiconductive devices. These subjects are reviewed in the last two chapters. Use of such particles is continuing and while such use is not necessarily relevant to current metal powder technology, which is emphasizing important mechanical applications, the availability of the sub-micron powders will have an influence on future developments.

In addition to the more sophisticated procedures for preparing finely divided metals, recent advances in ion-implantation have enlarged the scope of metal/polymer composites. Individual metallic ions are impinged upon specific areas of polymer or ceramic substrate to enhance a specific area with the characteristics of an applied metallic ion. This subject is reviewed briefly in this chapter. Ion implantation is relatively new and must be recognized as another technique for producing special metal/polymer composites.

The relative volumetric proportions of metal and polymer are influenced by the functional expectations of the end product, and the available processing equipment for fabricating the composite. In Chapter 3, there are examples of molding compounds where the dominating matrix is the polymer and the physical properties reflect the cured polymer structure. An equally significant processing development uses metal powders (particularly steel) as the dominating matrix and small amounts of organic polymers for bonding, lubricating, or controlling of very fine powders. These composites which consist primarily of steel powders are injection molded, and followed by high temperature sintering.[1]

When plastics materials comprise the major dominating matrix, the metals should be pre-alloyed to avoid separation of the alloy elements from one another in the preparation of metal/polymer molded composites. With two or more types of materials, gravity separation of materials of different density during handling and shipping must be considered. There are measures available to offset this phenomenon, such as the inclusion of finely divided fibers which will augment final properties. Similarly, consolidation of metal powders and plastic powders into rods, bars, or ribbons during thermal extrusion, followed by chopping into pellets will maintain uniformity of distribution. The individual pellets, typically 0.5 to 1.0 cm long, will represent discrete units of the desired metal/polymer combination and hence are adaptable for use in molding composites. When fibrous reinforcements are required, metal or ceramic fibers may be substituted for the metal powders as polymer fibers are more readily altered during high temperature processing.

Metal/polymer combinations, as well as metal/rubber combinations, have been known for some time. Some of the early composites were cited by Delmonte.[2] Continuing trends have been reported by Hausner over the years, in his reviews of powder metallurgy and forecasts in trade journals.[3] More current is Kamal's chapter on metal powders used in polymers.[4] Bhattacharya has edited a recent book on metal-filled polymers, for which this writer prepared a chapter on production technology for metallic fillers.[5] Poster's book on metallic powders should also be noted.[6] The American Society of Metals

(A.S.M.), Society of Manufacturing Engineers (S.M.E.), Society of Plastics Engineers (S.P.E.), Society for the Advancement of Material and Process Engineering (S.A.M.P.E.), Composite Group of Society of Plastics Industry (S.P.I.), Metal Powder Industry Federation (M.P.I.F.), Federation of Materials Institute (F.M.I.-England), and the American Society of Testing Materials (A.S.T.M.) have all shown an active interest in the promulgation of metal/polymer composites.

Recent years have witnessed a large influx of new high temperature resistant polymers, particularly in composite structures. Metallurgists have grown accustomed to low cost polyolefin polymers, polyvinyl chloride resins, acrylic, and polystyrene resins which make up the bulk of the 40 billion pounds that are produced annually in the United States. Their service temperatures are usually below 100°C. New high temperature resistant thermosetting and thermoplastic polymers, while more costly, have respectable performances up to 300°C. These will appear more frequently in future metal/polymer composites.

Among the newer polymers which will be used in metal/polymer composites because of their high temperature resistance are:

Polyphenylene sulfide (PPS)—Phillips Petroleum Corp.
"Xydar"—an aromatic polyester—Dartco Mfg.
"Ultem" polyetherimide resin—General Electric Co.
Polyimide ("Kapton") and PI2540—E. I. duPont de Nemours
Polyetherether Ketone (P.E.E.K.)—International Chemical Co.
Polyether sulfones—Union Carbide Corp.
Poly (p-phenylene benzothiazole) (PBT)—SRI and W.P.A.F.B.
Polyamideimide ("Torlon")—Amoco Chemical Co.
Polybenzimidazole—Celanese Corp.
Polyimide 2080—Dow Chemical Co.
Bismaleimides—Shell Chemical Co. and Ciba Geigy Co.
Silicones and Polytetrafluoroethylene (Heat resistant and flexible polymers)

Most polymers and metals can be formed into fine filaments and fine whiskers. Fibers are also being prepared from ceramic products such as aluminum oxide, silicon carbide, and boron (deposited on a tungsten filament), glass fibers, and fibers from organics such as polyaramids and high modulus polyethylene. Graphite fibers prepared from polyacrylonitrile and from bitumen pitch fibers are also commercially available. These fibers may be used as reinforcements for metal structures, conversely metal fibers and metallized organic fibers are used as reinforcements for polymer structures. Fibrous materials are contributing to new trends in metal/polymer composites.

PRODUCTION OF METAL POWDERS

The production of metal powders has been evaluated, and the properties of the powders has been reported by members of the Metal Powder Industries Federation (M.P.I.F.) and the American Powder Metallurgy Institute. Basic materials and testing specifications have been established for powdered metals by ASTM-B-9 committees. Steel powders are the most extensively used in the United States, with production of more than 250,000 tons of steel annually. This volume exceeds the powder production of all the other metal powders combined. Metal/polymer composites use many other alloys of aluminum, copper, nickel, and titanium, and developments continue.

In addition to fineness of particle size attained by methods of manufacture of metal and polymer powders, the shape of the powders influence the final composite and the chemistry of the surface. Some materials, and their preparation techniques, will favor spherical particles, while others offer irregular dimensions and clusters of particles. Manufacturers of these finely divided particles offer detailed descriptions or photomicrographs, as their selection will influence physical or electrical properties of the final composite part. Experiences with macro-sized aggregates suggest that a range of particle sizes rather than one specific size may enhance properties. Irregular metal surfaces are more receptive to compaction, while spherical or plate-like particles help achieve smooth surface coatings. Porous metal particles lend themselves to infiltration and impregnation by lubricants or polymers. Specialists in powder metallurgy should be consulted and their material and processing experiences solicited to achieve optimum results.

Various shapes of metal powders depend upon the methods of manufacture as described below. Descriptions are empirical, and while the sizes for spherical powders may be given with accurate dimensions, the largest overall dimension of the irregular shapes is frequently used as the size of the particles. Typical shapes and methods of manufacture are noted in Table 2-1. Information from commercial sources provide representative data. Generally, a broad range of sizes of metal powder may be made available by adjustments to the powder manufacturing equipment.

Conventional screen sizes for fine particles have been standardized for many years. Because commercial sources of finely divided particles use these screen sizes for specification purposes, Table 2-2, lists typical screen mesh sizes. The distribution of particle sizes usually centers on a limited size range, though small percentages above and below the nominal size are generally accepted.

Metals will appear in polymer composites not only as finely divided powders, but also as:

Thin metallic films, or vacuum coated metallized polymer foils which may be cut into small strips or chaff. The latter composites are used as anti-radar

Table 2-1. Typical Sizes and Shapes of Metallic Powders*

METAL	MANUFACTURING METHOD	REPRESENTATIVE COMMERCIAL SIZES AND SHAPES (SEE SCREEN SIEVES—TABLE 2-2)
Aluminum	Air atomized	100-325 Mesh—irregular and flake
Aluminum	Gas atomized	100-325 Mesh—spheroidal
Iron alloys	Chemical salt interactions	Sub-micron—spheroidal
Iron (99.7%)	Carbonyl decomposition	Spheroidal—5 micrometers
Iron (99.5%)	Electrolytic	Irregular—100 mesh
Copper	Electrolytic	Dendritic particles—100-325 mesh
Copper alloys	Various	Irregular—100 mesh
Nickel alloys	Atomization	Porous clusters—100 mesh
Nickel (99.8%)	Carbonyl decomposition	5 micrometers
Titanium	R.E.P. Atomization	Sponge, irregular 100-325 mesh

*Bulletins—Consolidated Astronautics, and Reade Co.

screens by the military. Epitaxial ultra-thin films used in semiconductors are described in Chapter 9.

Metal filaments or metal coated nonmetallic fibers, as continuous lengths or finely chopped to different aspect ratios (length to diameter). Metal filaments are also encased in extruded thermoplastics and chopped to fine pellets.

Flakes of metals such as aluminum, copper, and nickel offer effective barriers to moisture and also function in radiation shielding formulations.

Ion implantation involving metallic ions directed at specific areas is being used in electronic applications.

Amorphous, non-crystalline particles of metal result from the recent technology of very rapid solidification processes circumventing crystallization. Useful metallic-glass products are being developed as ribbons and fibers which are making many new contributions to metal/polymer composites.

Metal coated solid or hollow particles. Glass and other nonmetallics are now being coated with chemically deposited silver as a means for extending the more costly silver in conductive polymer adhesives and coating formulations. Silver coated spherical glass particles are typical.

Table 2-2. Screen Analysis of Metallic Powders (ASTM B 214-76)*

SCREEN SIZE	NOMINAL SCREEN OPENING (μM)	
No. 60	250 micrometers	Report percentage of
No. 80	180 micrometers	material by weight
No. 100	150 micrometers	remaining on screen and
No. 140	106 micrometers	percent by weight passing
No. 200	75 micrometers	through.
No. 325	45 micrometers	

*The Table is extensive and only a small part is reproduced here. Metal powders used in powder metallurgy are usually larger than 5 μm. Particles smaller than no. 325 mesh are considered "fines."

PRODUCTION OF FINELY DIVIDED METALS AND POLYMERS   17

Mixtures of particles of different sizes, amorphous or crystalline, fibers, and flat platelets may yield superior qualities to those obtained with a uniform size of metal powder. The demands of powder metallurgy necessitate clean, uncontaminated particles, while plastic matrices in molding compounds do not have such stringent requirements.

CHARACTERISTICS OF METAL POWDERS

There are several characteristics of metal powders which must be considered when making a selection. The overall sizes might be expected to fall in a range of 5 to 500 μm, though larger particles are available, and as indicated in the section on their preparation, metal powders are available in sub-micron sizes. An analysis of the metal powder characteristics from a powder metallurgy viewpoint, and from a plastics processing viewpoint are outlined below.

POWDER METALLURGY	CHARACTERISTIC	PLASTICS PROCESSING
Affected by particle shape and method of manufacture.	Porosity	In pressure molding the polymer will penetrate and impregnate porous fillers.
Molding and sintering achieve high strengths when theoretical density of the metal is approached by compacting under heat and pressure.	Apparent density	Thermal blending of components must accommodate volume of fillers and achieve compacted version of mixed ingredients.
Small amounts of ceramic contaminants, and absorbed gas layers adversely affect physical properties of final metal compacts.	Purity	Generally, not an important factor. Chemical consistency of product should be maintained.
Particle size and shape, density, and compaction in the mold cavity, shrinkage during sintering, have major influences, small size parts produced, "green" strength before sintering limits size.	Physical properties	Depends on a particular formulation and under uniform pressure and temperature, molded properties are consistent. Large parts up to several hundred pounds are produced.
Small amounts of polymer to improve the "green" strength before sintering and use of rapid plastic molding techniques.	←Basis for polymer additions—or—Metal additions→	Electrical property control to expand markets— improved heat resistance.

18 METAL/POLYMER COMPOSITES

The physical properties of powdered metal parts are considerably superior to those of molded polymer products. For the highest strengths, the laminated reinforced polymer structures described in Chapter 6 are superior. Because of emphasis on the properties of the polymer, a comparison chart shown below is reproduced from ASTM-B-310 listing typical tensile and compressive properties of sintered steel specimens, furnace cooled and quenched from 871°C (1600°F). The carbon contents of Class A, B, and C materials are typical of steel specimens. The influence of apparent density of the steel powders shows up in the final properties of the sintered specimens. See Table 2-3.

Table 2-3. Typical Properties of Carbon Steel—ASTM-B-310[7]

	CLASS	DENSITY RANGE, G/CM³	ULTIMATE TENSILE STRENGTH				ELONGATION IN 1 IN. OR 25.4 MM, % AS-SINTERED	YIELD POINT IN COMPRESSION, 0.1% OFFSET AS-SINTERED	
			AS SINTERED		QUENCHED AND TEMPERED				
			PSI	MPA	PSI	MPA		PSI	MPA
I	A (C = 0.3 max)	5.9	16 000	110	...	...	2.0	12 000	83
	B (C = 0.3 to 0.6)	5.9	20 000	138	30 000	207	1.5	18 000	124
	C (C = 0.6 to 0.9)	5.9	26 000	179	40 000	276	0.5	22 000	152
II	A (C = 0.3 max)	6.3	20 000	138	...	...	3.0	17 000	117
	B (C = 0.3 to 0.6)	6.3	26 000	179	40 000	276	2.0	22 000	152
	C (C = 0.6 to 0.9)	6.3	34 000	234	50 000	345	0.5	26 000	179
III	A (C = 0.3 max)	6.7	26 000	179	...	...	5.0	20 000	138
	B (C = 0.3 to 0.6)	6.7	34 000	234	50 000	345	3.0	24 000	165
	C (C = 0.6 to 0.9)	6.7	44 000	304	64 000	442	1.0	28 000	193
IV	A (C = 0.3 max)	7.1	30 000	207	...	...	8.0	25 000	173
	B (C = 0.3 to 0.6)	7.1	50 000	345	70 000	482	4.0	35 000	241
	C (C = 0.6 to 0.9)	7.1	60 000	414	80 000	552	2.0	42 000	290
V	A (C = 0.3 max)	7.2 min	40 000	276	...	...	10.0	30 000	207
	B (C = 0.3 to 0.6)	7.2 min	60 000	414	80 000	552	5.0	42 000	290

MANUFACTURING PROCESSES FOR METAL POWDERS

Mechanical Attrition. Grinding and breaking up of the more brittle metals takes place when materials are charged into ball mills and impact hammer mills. The hard, crystalline metals and their alloys may require several stages for grinding to fine particles. Screening and air classification will separate and size the particles. Rough fracture edges would be expected, except in the case of prolonged ball milling with ceramic balls. While hardness has much bearing on the ease of grinding any material, the following factors are significant: crystalline structure, friability, toughness, and fibrousness. Less energy is expended in wet grinding as compared to dry grinding. For a review of equipment developed for grinding a wide variety of materials see Reference 8. Many devices have been developed for handling non-metallic materials as well as metallic materials, and the experiences of one industry are extended from one industry to another. Recent developments in manufacturing processes for producing metal powders are described below and summarized in Table 2-4.

Atomization. Stainless steel and other alloy powders are produced by atomization. Streams of the molten metal are broken up by hot gases or hot liquids impinging upon the material. High yields of 100 mesh powder are generally reported. Argon gas atomization, accompanied by gas recirculation and purification, yields powder particles with an oxygen level less than 100 parts per million (ppm). Argon gas atomization is also being used in the production of powdered high speed tool steel alloys (Fig. 2-1). General Electric patents describe closely coupled nozzles, located less than 1.2 centimeters from molten metal surfaces to atomize the metal into fine particles.[9]

Molten aluminum metal is drawn through a tube that is lined with refractory material to protect the tube from erosive effects. A jet of hot air impinges upon the aluminum issuing from an orifice, blasting the molten metal into fine particles. The aluminum powder prepared by atomization processes is used as the starting material for the production of aluminum flakes which are used extensively in pastes and coatings. The particles and flakes are covered with a film of oxide which comprises about 0.1 to 0.2% by weight. The flakes are obtained by the impact of polished steel balls in a rotating mill, while the flakes are wet with a liquid hydrocarbon (which prevents explosion hazards). A small amount of stearic acid lubricant is added to facilitate leafing action of aluminum flakes when they are employed in paint vehicles. This process is called the "Hall" Process after its inventor at the Aluminum Company of America.[10]

At the end of the operation, as much of the liquid as is practical is removed. The aluminum "cake" is either dried in vacuum to produce dry flakes, or more frequently, it is adjusted with mineral spirits or lacquer solvents with specified metal content, usually 60 to 70%. The finest flake aluminum averages less than one micrometer in thickness.

Table 2-4. Methods for Manufacture of Metal Powders

Mechanical attrition	Best for brittle particles and dendritic crystalline shapes—crushers, grinders, ball mills, hammer mills. Some metals such as tin, are ductile at room temperature and must be heated to be brittle before grinding.
Chemical reduction of metal oxides	Powdered metallic oxides are reduced by gases such as hydrogen or methane. In the case of magnetite (Fe_3O_4), charcoal is used to produce a low cost iron powder.
Decomposition of metal carbonyls	Accomplished on iron and nickel carbonyls in presence of carbon monoxide at high pressure and temperatures. Metal carbonyls are highly poisonous liquids. Fine spheroidal particles are formed.
Atomization	Atomization processes are available for the fragmentation of low melting and ductile metals such as lead, tin, zinc, and aluminum. The molten metal (usually below 700°C) is blown through a nozzle by a gas preheated above the melting point. Inert gases are used to avoid oxidation and minimize pyrophoric tendencies. Water and vacuum atomization are also used.
Electrolytic processes	Through the use of appropriate electrolytes, current density, and special additives, very fine powders may be produced. Brittle cathodes which may be formed may be ground to a powder.
Precipitation from solution	Less noble or cathodic metals in powder form, e.g., copper, iron, or nickel, may be precipitated from their solutions by aluminum powder. The precipitated metal takes on the shape of the precipitator metal.

The rapid quenching of atomized liquid aluminum alloys in a nonoxidizing atmosphere results in a fine dispersion of particles, which are used in the manufacture of powder metal.[11] Rapid induction heating of the metal powder compacts is also cited in Reference 2-11.

Ultrasonic Gas Atomization. Using the Swedish-Kohlswa process, fine spheroidal-like particles are produced by high quench rates of about 10^5 °K/second.

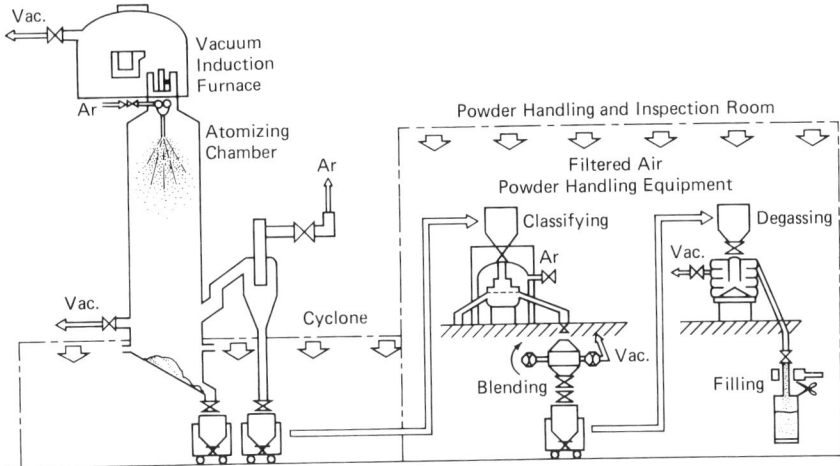

Figure 2-1. Argon gas atomization equipment. Melt nozzle diameter is 5 mm and gas pressure is 40 Kg/cm^2. Kobe Steel Co.

The process depends upon the shearing action of high velocity gas (about 2 Mach) with pulsed frequencies.[12] The powder size distribution is attributed to the spread of the atomizing gas jets during travel from the nozzle exit to the metal stream. Aluminum alloys and iron-silicon powders were studied in this manner.[13]

Centrifugal Atomization. Molten metal is tangentially ejected as droplets from a rapidly spinning ceramic crucible or disk, etc. The droplets solidify into particles while flying away in a gas or vacuum. The cooling rate can be accelerated above 10^5 °K/second by using a cooling medium with high thermal conductivity. Particle size will be greatly dependent upon the peripheral velocity of the spinning body. Figure 2-2 illustrates schematically a process for inert gas atomization and rotating electrode processing (REP). The centrifugal atomization equipment of the Daido Steel Company of Japan is shown in Figure 2-2. Illustrations are adapted from Reference 2-12.

High Intensity Arc and Plasma Rotating Electrodes Process (PREP). High intensity arcs have been used to produce fine metal powders in a controlled inert atmosphere or vacuum. The feed material is made the anode and the metal is rapidly melted and evaporated.[14] Rotation electrode processes, including a helium plasma arc, contribute to the centrifugal atomization of titanium and superalloy powders (Fig. 2-3). The consumable electrode forms the anode, as

22 METAL/POLYMER COMPOSITES

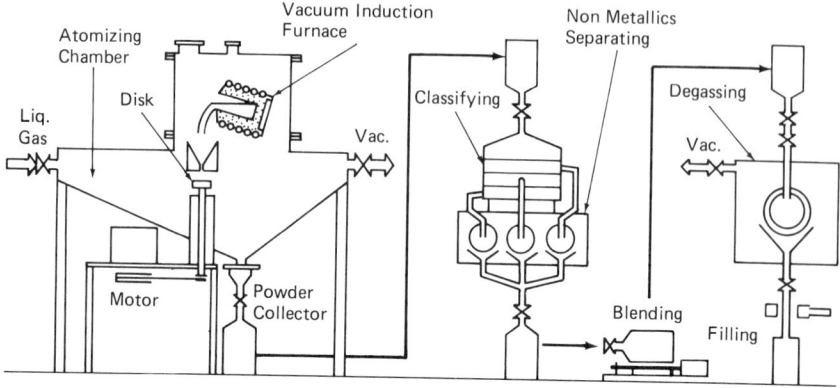

Figure 2-2. Centrifugal atomization equipment. Daido Steel Co. Ltd.

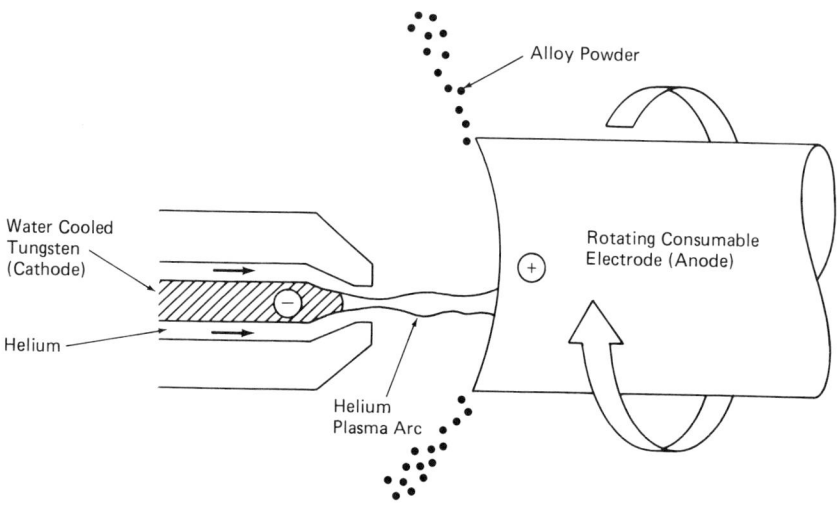

Figure 2-3. The plasma rotating electrode process produces clean powders of cobalt alloys and titanium alloys.[3]

the arc is struck from the tungsten cathode. High velocity inert gases carry the molten metal away from contamination with the tungsten.

Decomposition of Metal Carbonyls. This technique afford further processes for obtaining fine metal powders of uniform sphericity and high purity. The reaction of iron with carbon monoxide may be used to prepare iron pentacarbonyl Fe(CO)$_5$. It is a yellow to dark red colored liquid, which has a specific gravity of 1.453 and boiling point of 103°C (highly toxic).

Ultrafine Magnesium Powder. A fine powder of magnesium, with particle size of less than one micrometer, was prepared by vaporizing magnesium in an inert gas such as helium at 10^3 Pa pressure.[15] Infrared lamps near the surface accelerate the mixing of vapor and the inert gas before cooling.

Electrodeposition of Iron Powders. Iron powders were obtained from electrodeposition on a rotating nickel cathode immersed in a FeCl$_2$ aqueous solution. The current was 1.8 ka/m^2 at temperatures of 318 ± 2 °K. A solution of epoxy resin functioned as an upper layer in the bath, inhibiting oxide formation.[16]

Zinc Powder Preparation. Zinc powder is prepared by atomizing molten zinc with a hot gas jet at 350°C. Spherical particles were obtained at 100-200 screen mesh size.[17] Quantities of zinc powder are collected in condensers during distillation.

Fine Copper Powder. A Japanese patent describes the preparation of less than 0.1 μm powder, about 400 angstroms in size with a 96% yield and a surface area of 16 m^2/gram. Details show a solution of copper sulfate, 0.5 mol (liter), at a pH of 6.0 at 20°C. To this is added aqueous NaBH$_4$, developing a fine copper slurry.[18]

Iron Alloy Acicular Particles. Iron alloy particles for magnetic purposes, which include up to 10% copper or manganese, are cast in a melt of a base metal of copper or gold and a strip prepared by rapid solidification (melt spinning). The base metal is electrolytically dissolved to recover the special iron alloy particles.[21] This is a costly procedure, and other alternatives are described in the chapter on magnetic applications.

The measures for preparing composites may in themselves reduce the metals to fine powders or finely divided particles. A few of the measures are briefly noted in the following examples, which indicate materials for molding purposes. These techniques are discussed in more detail in Chapter 3. The chopped metal fibers will add impact strength and stiffness to polymer compounds, and for magnetic recording tapes, the needle shape has special significance.

Nonmetallic Fibers Coated with Metal. Carbon, graphite fibers, and glass fibers are surface treated with titanate coupling agents and coated with a thin film of metal. These are positioned within a thermoplastic jacket by extrusion. The entire assembly of thermoplastic polymer and metal coated fibers are chopped into pellets measuring two millimeters in diameter and five millimeters in length. These metal/polymer composite pellets have been used by Toshiba in molding.[20]

Nickel Plated Graphite Particles. Tows of continuous graphite fibers (12,000 count) were electroplated with nickel and dipped into a solution of polyethyloxazolene. The impregnated fibers were heated at 160°C to remove solvent and then chopped to $1/16$ to $1/4$ inch particles containing a large volume of fibers. The particles were injection molded with bisphenol-polycarbonates to form electromagnetic shielding.[22]

Improved Thermal Transfer Recording Material. Powdered copper (1.5 μm) was spread on a polyethylene terephthalate film and irradiated at 10 MHz for 10 seconds. The film was pressure rolled to force the copper particles into the film. Electrical power energy transfer requirements were about 30% less than in the absence of copper because of improved heat transfer.[23]

Stainless Steel Fiber Molding Compounds. Stainless steel fibers are melt extruded with a thermoplastic acrylonitrile-butadiene-styrene (ABS) resin system. The composite is pelletized to make available molding compounds suitable for electromagnetic wave shields.[27] Recent developments have produced more efficient use of stainless steel wires in molding compounds by extruding a bundle of metallic wires, coated and insulated with the insulative polymer. See Figure 2-4 which illustrates the bundle of wires (1) contained in a plastic sheath (2). These extruded rods with oriented encapsulated wires are chopped into small pellets suitable for inclusion with a compatible molding compound.[28]

Chopped stainless steel wires are used in wool and nylon carpets to function as a static eliminator. A competitor of this technique is the anti-static nylon filaments introduced during tufting—a process of the BASF Corporation.

Plastics, ceramics and metals are available in fiber form and in some instances as very fine whiskers. In the creation of reinforced plastics, fiber reinforcements of polymers, ceramics, and metals are used to impart added strength and stiffness. There are increasing activities in the development of

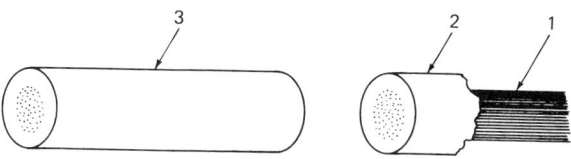

Figure 2-4. Schematic view of 1. steel fibers, 2. extruded polymer sheath, 3. typical molding pellet.[28]

fiber reinforced metals, particularly in the use of inorganic fibers such as silicon carbide and aluminum oxide.

In the following paragraphs are noted additional examples of activities which use metal fibers for developments in metal/polymers composites.

Fiber Reinforced Metals. Aluminum wire reinforced with silicon carbide fibers, has been chopped to short lengths of the order of one centimeter or less, and used as reinforcement for plastic molding compounds. Long lengths serve as structural reinforcement for polymer laminates. Added stiffness (higher moduli) and the strength of these silicon carbide fiber reinforced metal wires pass these qualities on to the final composite.[19]

Metal Fiber Reinforced Metal. Tungsten wire fibers 120 μm in diameter have been used as reinforcement for copper alloys. The tungsten wires were cleaned with NaOH solution, rinsed in distilled water and inserted into a closed end ceramic tube. Seventy-five to 80% volume of tungsten fibers were embedded in copper alloy matrices with the following results[24]:

MATRIX	AVERAGE TENSILE STRENGTH (PSI)
Tungsten Wires	330,000
Pure Copper	240,000
Copper Nickel (.33%)	235,000
Copper Aluminum (2.6%)	130,000
Copper-Niobium (1.0%)	230,000

Boron fibers fulfill important roles in reinforced laminated structures for aerospace. Boron is deposited upon a fine tungsten wire in an inert (argon) atmosphere from a B_4C vapor (evaporated at 1900°C). The boron fibers comprised a tungsten core of 12 μm and an external diameter of 100 μm. Tensile strengths of 300,000 psi (2.1 GPa) and elastic modulus of 50×10^6 psi (343 GPa) are obtained. For added moduli and added tensile strength, they are very useful to the composite. The use of continuous boron fibers in metal matrix composites (MMC) has been effective and is discussed in detail in Reference 2-37.

Whiskers. Fine metallic and nonmetallic whiskers have demonstrated remarkable tensile strength and high modulus of elasticity. Figure 2-5 illustrates the characteristics of fine iron whiskers. The rapid increase in tensile strength over 1×10^6 psi is noted at 1 to 2 μm in diameter.[25] The whiskers are frequently in the sub-micron range in size, and tend to have a high degree of crystal perfection. Silicon carbide whiskers were first produced commercially

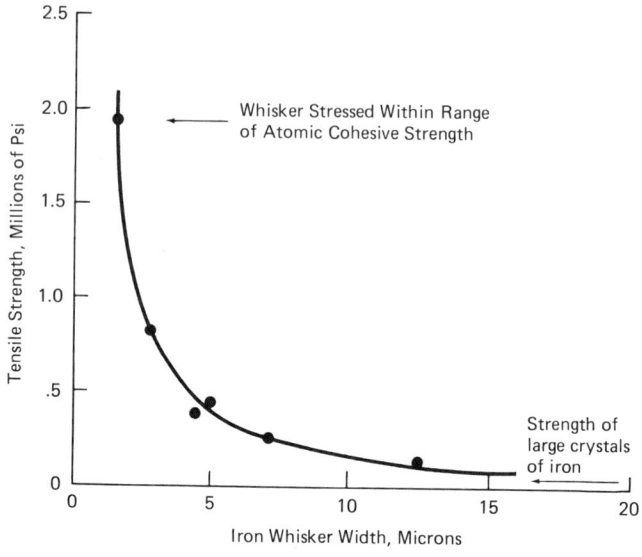

Figure 2-5. Properties of iron alloy whiskers.[25]

in 1965 by Carborundum Co. with an initial diameter of 1 to 5 μm and lengths of 30 to 100 μm. Tensile strengths of 3×10^6 psi and 70×10^6 psi elastic modulus were reported. About the same time (1964) Brunswick Co. was producing fine metal fibers as small as one micrometer from 304 type stainless steel. Single filaments of stainless (12 μm diameter) showed tensile strength of 275,000 psi and elastic modulus of 29×10^6 psi. The testing of fine fibers requires careful techniques, and while the tensile strength of a single fiber may be high, in a composite of larger dimension, strengths are substantially lower.

Milewski's pioneering efforts in the use of whiskers for reinforced plastics was reported in 1964.[26] Figure 2-5 illustrates a relationship between whisker diameter and the tensile strength of pure iron. Similar relationships were established on fibers drawn from sapphire mineral ($Al_2 O_3$). At the present time commercial uses of metal and inorganic whiskers are limited, however, the technical advantages are so great that it is safe to predict greatly increased use in the future of sub-micron metal whiskers and inorganic whiskers in composites.

PRODUCTION OF POWDERS FROM POLYMERS

Finely divided powders of polymers require different processes than those used for metals. With the exception of naturally occurring resins and polymers, the synthetic resin polymers of the past 100 years are prepared from monomers and chemicals capable of polymerization reactions. The monomers are basic chemical molecules capable of polymerization to form high molecular weight polymers. Polymerization processes may take place as follows:

- *Bulk or Mass Polymerization.* Liquid resins from polyesters, epoxies, urethanes, alkyds, melamine formaldehyde, or phenol formaldehyde, will cure as they polymerize either in bulk or as an impregnant. Phenolic and melamine resins are partially polymerized (called "B" stage) and ground up before being blended with organic, inorganic, or metallic fillers. Further processing during molding converts the product to an infusible state. Thermoplastic monomers are also cast as liquids, and cured during bulk polymerization in molds. Sheets of acrylic, for example, are cast in this manner. Bulk polymerized thermoplastics may be ground, as noted below.
- *Solution Polymerization.* This process is adaptable to other potential thermoplastic or thermosetting forming materials in an appropriate compatible solvent. These products are used in coatings, impregnants, and adhesives with a large variety of fillers. The technique is not usually associated with powder technology.
- *Suspension Polymerization.* Very finely divided polymer particles may be obtained from colloidal polymerization in which the initial thermoplastic monomers are processed in nonsolvents to very fine suspensions before the polymerization reaction proceeds. Polyvinyl chloride polymers, for example, are prepared in this manner.

High temperature resistant polymers such as polytetrafluoroethylene ("Teflon," for example) are not easily prepared by thermal extrusion processes because of their high melt viscosity. Water based dispersions have provided powders as small as 0.2 μm. The coating of PTFE on metal cookware is the basis of a major industry in the United States.

Low Temperature Grinding. Solid masses of thermoplastic polymers or partially polymerized thermosetting polymers may be comminuted further by mechanical attrition in machines similar to those used for fracturing metals. Because of the toughness of some of the thermoplastic polymers, low temperatures (achieved with the aid of solid carbon dioxide or liquid nitrogen) make the grinding process practical. At these temperatures the polymers become brittle and lend themselves to grinding. Specialized equipment is available for the low temperature grinding of polymers on a commercial basis. Good powder flow of polyethylene is enhanced by larger particle sizes, narrow size distribution, and high sphericity.

COMBINING FINELY DIVIDED METALS AND POLYMERS

Finely divided metals and their alloys are blended with polymers and their plastics in many ways. Low molecular weight polymers may be liquids, semi-solids, or they may be solid particles. To insure the identities and uniformity of the metal components when blending metals and polymers, the metals will be pre-alloyed before being pulverized, and similarly polymers will be cured to a particular molecular weight range and pre-compounded into plastics before being blended into the metal/polymer composites. It should also be noted that pre-compounded plastics may have had fillers, fibers, pigments, flow agents, or cure agents incorporated into their composition before being blended into the metal/polymer system.[35, 36] In following chapters, the fabrication by molding, casting, spraying, or impregnation will be reviewed for blended combinations.

Processing techniques for blending together finely divided plastics and metal alloys may require several methods or combinations some of which are listed below. More detailed descriptions follow.

Thermal Blending. Under intensive mixing and high heat, the polymers are melted and brought into intimate contact with the metal powders. The use of three roll mills and Banbury mixing equipment (developed earlier for the rubber industry) has been followed in recent years by extrusion mixing of the compounds.

Dry Blending. Tumbling of dry polymer particles and fine metal particles can be done at low cost. Size reduction may also be accomplished at the same time.

Solution Blending. Liquid polymers or solutions of solid polymers permit batch mixing of metal particles or fibers with polymers.

Aqueous Dispersions. The use of equipment designed for the paper and pulp industry permit blending of highly diverse metals, fine polymer particles (or colloids), ceramics, and fibers.

Ion Implantation. The new techniques of ion-implantation permit the introduction of metallic ions into specific areas of polymers or ceramic substrates.

Thermal Blending. Combinations of solid particles of plastics and of metals are prepared by the use of equipment which will mix together two or more different materials into a uniform single homogeneous mass. Historically, three roll mills and Banbury mixers, used by the rubber industry, were adapted for the preparation of particulate filled thermosetting polymers. They are still used for short run trials in the laboratory for many types of polymer blends because of their comparative ease of cleaning and maintenance. More recently, screw extruders have attained a high degree of sophistication because of the multiple controls available for optimizing the quality of the extruded composite mass. Temperatures are attained which melt the polymer, adhesively bonding it to the solid metallic particles.

A modern twin-screw extruder is illustrated in Figure 2-6 which highlights some of the features that are needed by manufacturers requiring an output up to 5,000 kg of compound per hour. This equipment is usually employed for metal/polymer molding compounds. The output generally passes through a dicing chopper or pelletizer which delivers the material in a form suitable for the feeding hoppers of molding machines. High performance "engineering type" thermoplastics utilize such equipment.

Industry utilizes two broad categories of compounding systems—batch (or discontinuous) systems and continuous systems. It is common practice to blend several batches in a suitable tumble blender, thus uniformly interdispersing the ingredients to achieve a uniform distribution of composition throughout the mass. In a continuous system, the raw materials are fed into a twin-screw extruder, insuring a continuous flow of compound at the end of the machine. The necessity for controlled introduction of diverse raw materials, and the importance of maintaining close temperature controls at the various stages, are apparent. Inorganic fillers and metal particulates will produce wear on the screws and inside of the barrel of the screw-extrusion machine. There are no exact formulas to predict the amount of wear, though the necessity for continuous quality control and inspection of the final output, are essential to monitor the useful working life of the equipment. Economics necessitates the computation of replacement costs, which must be factored into the selling price of the metal/polymer composite material.

Design features of the twin-screw elements for melting and mixing are illustrated in Figure 2-7. It depicts differences in the pitch and depth of screws which control kneading and shear actions. The twin-screw extruder increases the range of conveying capability above that of a single screw. Downstream feeding into the melt at any stage can be carried out if desired. Wear in the equipment will be at a slow rate, and must be correlated with minor property changes of the compounded material. Operational adjustments in temperature and feed-through time will extend the useful working life of the equipment.

While the sophisticated twin-screw equipment has been engineered for efficient through-put of metal/polymer composites, the single screw extruders have the advantages of good versatility for colorants, fillers, and additives which

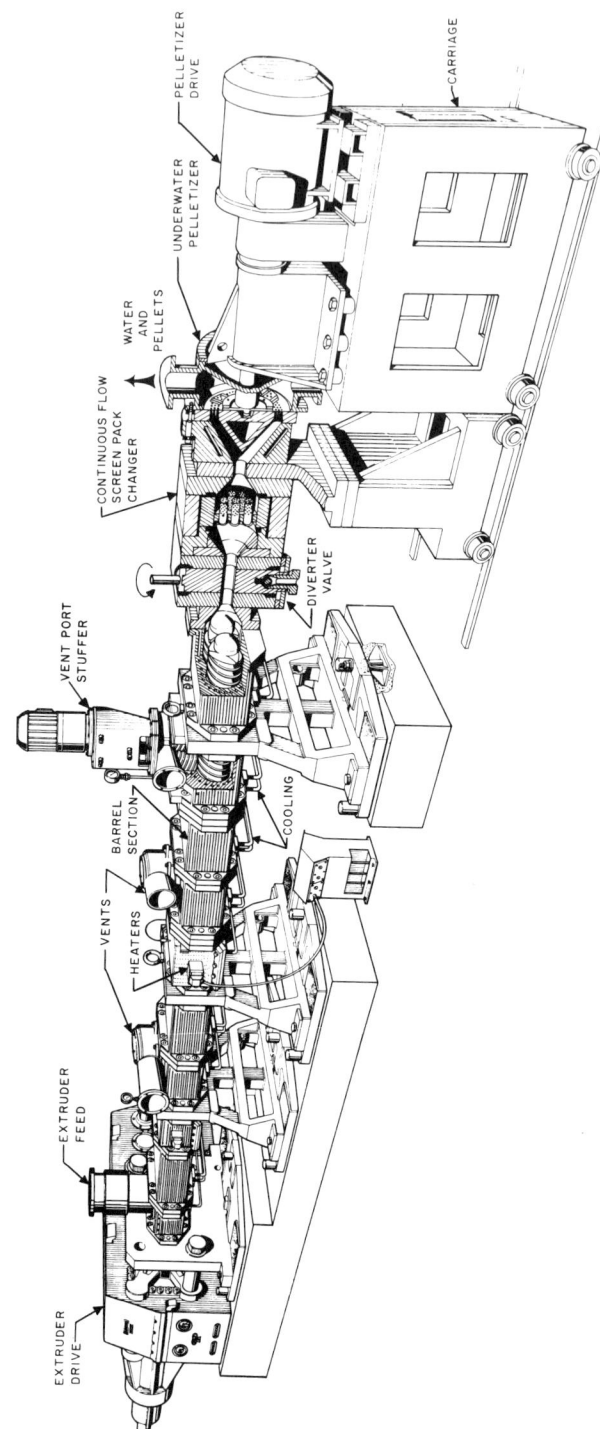

Figure 2-6. Plastics compounding equipment. Courtesy of Werner-Pfleiderer.

PRODUCTION OF FINELY DIVIDED METALS AND POLYMERS 31

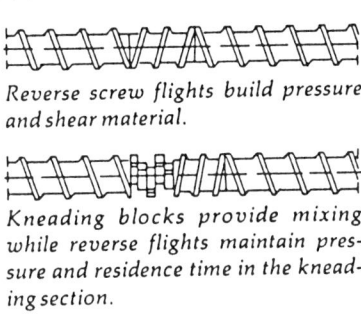

Reverse screw flights build pressure and shear material.

Kneading blocks provide mixing while reverse flights maintain pressure and residence time in the kneading section.

Kneading blocks, used alone, are suitable for less difficult mixing applications.

High-pitch left-handed element provides high shear levels.

Figure 2-7. Examples of twin-screw elements for melting and mixing. Courtesy of Plastics Compounding[33] and Werner-Pfleiderer.

require less maintenance, also they are more easily cleaned, and are simpler to operate.[33] Screw extrusion compounding is adaptable to metal/polymer blends.

Hydrostatic extrusion is practiced extensively for the preparation of metallic wires and rods. High deformation processes are involved in reducing a hot metal billet (aluminum and copper) or a mass of molten polymer. The ratio of the cross section of the billet to the cross-sectional area of the extruded product influences physical properties. In a later chapter on the preparation of structural elements of metal wire reinforcements in both polymers and in metal matrix, extrusion will be examined as a technique for preparing continuous components of metal/polymer composites.

The energy expended during thermal blending by screw extrusion is large. Todd has summarized some of the key heat transfer and heat generation features in a few basic equations.[32] The polymer begins to melt because of the friction caused by particles being forced against one another by the rotating screws, and by heat transfer through the barrel walls.

Heat input into blend = $WC_p \triangle T$
 W = Mass Flow Rate—kg/Hour
 C_p = Heat Capacity—kJ/kg (Sum of metal and plastic)

32 METAL/POLYMER COMPOSITES

T = Temperature rise in °K
Work energy imparted by extruder
$= W \triangle P / Ep$
$\triangle P$ = Pressure Differential (Pa or N (m²))
W = Mass Flow Rate—kg/Hour
E = Extruder Efficiency Factor (0.1 is typical)
p = Melt Density—kg/m³

Industry usually expresses the degree or intensity of mixing as horsepower per kilogram. Low intensity mixers operate at a ratio of 1 horsepower (hp) per 75 to 150 kilograms of material, medium intensity mixers operate at 1 hp per 10 to 15 kg of material, and high intensity mixers operate at 1 hp for 1 to 2 kg of material. These intensity classifications are also extended to continuous and batch compounders for blending and mixing of liquids and pastes. See the summary in Table 2-5. Though temperature control is still a factor in liquid mixing, it is not the major influence as it is in thermal blending of dry solids by screw extrusion equipment.

Before leaving the subject of thermal blending of polymer/metal composites, some of the earlier practices of mixing inert fillers into polymers should be mentioned. Most of these techniques were developed for mineral fillers such as mica, glass fibers, carbides, pigments, oxides, carbonates, asbestos; to a limited extent, metal powders; and many organic fillers such as cellulose, wood fibers, and carbonaceous products. Several good references on fillers are available on this subject.[35, 36]

On a heated two-roll rotating mill, for example, the amount of shear on the material being compounded is dependent upon the nip through which the material passes. The same action occurs in screw compounding where the screw flights carry the material by friction towards the point where the two screws meet, forming a bank of material. Blending and cross blending of filler/polymer combinations will take place on the heated rolls. Blends are removed after complete mixing of all ingredients takes place. The blended material is removed, and cooled, and subsequently broken up in hammer mills or attrition mills, and screened to a desired particle size for molding purposes.

When thermosetting resins are involved the time and temperature parameters on heated rolls are limited to avoid precure of the resin, multiple batches are obviously involved, and the blending of several batches in a rotating tumbler or ball mill reduces batch to batch variations. The more modern screw extrusion blenders are designed to overcome these limitations.

Dry Blending of Powders. Powder coatings of polymers and metals have been developed for covering electrical components such as capacitors, coils, small transformers, resistors, and assemblies which utilize metal frames, shells, and steel laminations. Finely divided plastics or composites of metals and polymers, which may have been prepared by thermal extrusion blending and then classified in particle size, may be applied and fused on thermally heated metal substrates. Application would take place in a fluid aerated bed or electrostatically sprayed on components. Further details are discussed in Chapter 5.

Solution Blending. The oldest technique established for the blending of finely divided metals into plastics matrices is accomplished by dissolving some of the polymer in effective solvents. Polyolefins such as polyethylene are not readily soluble and rely on aqueous dispersions for compounding. The resulting solution should be low enough in viscosity to permit supplementary materials such as flow control agents, coloring agents, inert fillers, chopped fibers, cure promoters, etc., to be effectively distributed in the whole mixture. The addition of the finely divided metals, which have a higher specific gravity than the solvated polymer, is usually accompanied by thixotropic agents (such as very finely divided fibers and/or colloidal silicas or clays) to minimize the rate of settling of the metal powders. Metal powder filled sprayable paints may require re-suspension before application, as they may settle in their containers during storage. Flaked metals such as aluminum are frequently used in solutions of metal/polymer composites and would be processed like other particulates. Finely divided magnetic particles are dispersed in polymer solutions for application to polymer films in the manufacture of magnetic recording tapes.

Solvent recovery may become a significant factor and appropriate chemical engineering practices will be necessary for efficient recovery. One should also be cognizant of the fact that removing solvents may also strip off low molecular weight portions of the polymers and alter properties of the coated materials.

If the intent of the solution mixing is to produce an intimate physical blend of all ingredients for use as a dry composite mixture, some tumble drying and vacuum drying is required to remove the last traces of solvent, and to break up undesired agglomerations of the metal particles and the polymers. Application of the combination of finely divided metals and polymers as a surface coating or a trowelable paste or adhesive putty, will reveal that the retention of at least a portion of the solvent is helpful during application.

Mixers and blenders are available for dispersing metallic particles in liquid polymers or solution dispersions of polymers. Many types of apparatus are available and Reference 33 illustrates various classes. Selection of equipment is influenced by the viscosity of the composite mix of inert particles and resin binder, as well as the batch size. A large choice is available from heavy-duty double-planetary mixers for semi-pastes, to high-speed high-shear dispersers for more fluid blends under 20,000 centipoises in viscosity at 25°C. Figure 2–8 illustrates commercial equipment with typical rotating blades that are available for high speed dispersers. The thoroughness of the blending can be as-

34 METAL/POLYMER COMPOSITES

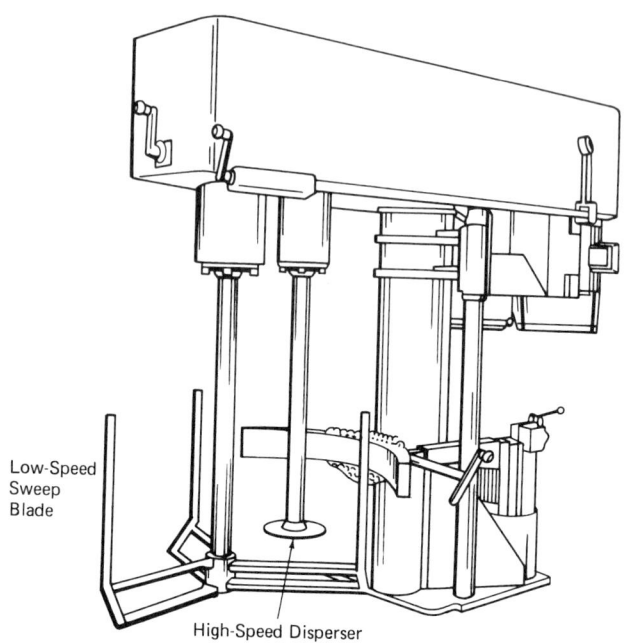

Figure 2-8. Dual-shaft high-shear, high-speed disperser with sweep blades. Courtesy of *Plastics Compounding* and Myers Engineering, Edgell Pub. Denver, CO. 1988.

certained by closely following changes in viscosity and the prevailing temperature.

For the high temperature thermoplastic polymers listed earlier in this chapter, costly solvent systems may be necessary. The ease of melting low density polyethylene permits compounding with particles at more modest temperatures. In Chapter 5 examples will be presented which deal with techniques of applying polymer coatings on metal structures.

Dispersions of thermoplastics in plasticizers, which at room temperature do not fully solubilize the polymer, are known as plastisols. These are also discussed in Chapter 5.

Table 2-5 summarizes typical mixing and blending equipment which do not require the high pressures and temperatures needed for thermal blending.

Aqueous Dispersions. There are advantages to the use of aqueous dispersions for compounding purposes, particularly for the introduction of a large volume of randomly distributed fibers such as glass, carbon, ceramic, metal, and metal coated polymer filaments. Unlike thermal blending or the intensive

Table 2-5. Mixing and Blending Equipment

DESCRIPTION	TYPICAL FUNCTIONS
Ribbon blenders, conical screws, tumbling barrels and large drums—some with porcelain balls for grinding.	Dry powders and low density plastics of diverse materials. Pre-blending of compounds before intensive mixing in screw extruders or heated rolls. Helical ribbons of steel are attached to rotating shafts, enclosed in a suitable trough. Double cone blenders will hasten commingling of components of powders, pellets, and fibers.
Medium intensity mixers—nonfluxing, planetary mixers.	Suitable for metal/polymer mixtures. Mixing blade actions and paddles promote good dispersion. PVC plastisols, polymer premixes, polyolefins, etc. will benefit from heavy duty planetary mixers and special sigma blade designs. Pastes and high viscosity liquids have been accommodated.
Low viscosity, high speed, high shear disperser with sweep blades.	Suitable for dispersions in non-solvent systems. Break-up of agglomerates and fragmentation of particles. Special serrated disk blades will create mixing vortices at high speed. Adaptable to PVC plastisols, polyesters, and polyolefins with or without finely powdered metals.

mixing of pastes, the tendency for fragmentation of the fibers is reduced in the presence of a large quantity of water. High speed mixers and rotating propellers will achieve good mixtures of potential composites.

Controlled amounts of the diverse materials are added in a carefully planned sequence for optimum results. The different materials to be combined should not be water soluble or chemically reactive with water, though during emulsification there are surface active agents which control the effectiveness of material distribution. Alloying ingredients which are an option for metals, or flow control agents which are an option for polymers, may be added in small amounts, depending upon the vortex velocity of the water and the action of the mixing blades to achieve distribution. The unusual combinations of fibers, metals, polymers, and/or ceramics are finally deposited upon fine screens (much like those used in paper and pulp manufacturing). Excess water is drawn off, leaving a coherent moist pad of thoroughly blended materials, which must be dried and processed at higher temperatures and pressures. Extra expenses are involved in removing excess water, but if conventional thermal blending of large masses of metals and plastics have proven unsuitable, the use of water emulsions should be considered to prepare metal/polymer composites.

Suspension polymerization is another aqueous dispersion technique for preparing fine particles of polyvinyl chloride. The suspension process is a batch process in which droplets of vinyl chloride monomer are suspended in vigorously agitated water. The monomer droplets are stabilized by suspension aids such as polyvinyl-alcohol or water soluble cellulosic esters. The polymer is formed by a free radical mechanism, generated from monomer soluble peroxides. Particle size and particle distribution of the resin are controlled by the degree of agitation and the type of suspension aid. It is conceivable that metal particles could be added after polymerization is complete.

Ion Implantation. Developments in the field of ion implantation make possible controlled modifications of the physical, chemical, and electrical properties of diverse substrates of metals, polymers, and ceramics. The characteristics developed on the selected surface area (or slightly underneath) by ion implantation are different from the characteristics of the bulk material. Hence it is technically possible to have multiple and diverse characteristics on the same substrate. Ion implementation is finding a special role in microelectronic devices, a subject that is examined in greater detail in Chapter 9, where metal/polymer composites in sub-micron dimensions are reviewed.

It is now possible to develop properties, which are independent of the bulk characteristics on the surface of a material. A stream of ions is directed towards selected areas of a substrate. The ions stop a short distance below the surface because of collisions with the target atoms. Concentrations of 10^{14} to 10^{21} ions per cubic centimeter are typical. Because ion beams can be produced from virtually any element, these ions are injected into the near-surface regions of a solid at concentrations well above the expected solubility limit. This method produces metal/plastic distributions that are not necessarily uniform, but highly specific in certain limited areas. These products differ from natural distributions on the Earth, where the process of erosion and geological activity have produced a complex commingling of minerals. The modern technology of ion implantation can be very specific in element distribution.

As illustrated in Figure 2-9, a gas is introduced into a chamber where it is ionized by electrons emitted from a hot filament. Metallic elements require more sophistication, and may require the decomposition of a volatile compound. Positive ions are extracted from the ion source and directed toward a mass separating magnet which selects a species for impingement on the substrate (Ref. 2-34).

Figure 2-10 adapted from Reference 2-29 illustrates how the ions lose their energy: by excitation and ionization of electrons and by elastic collisions with

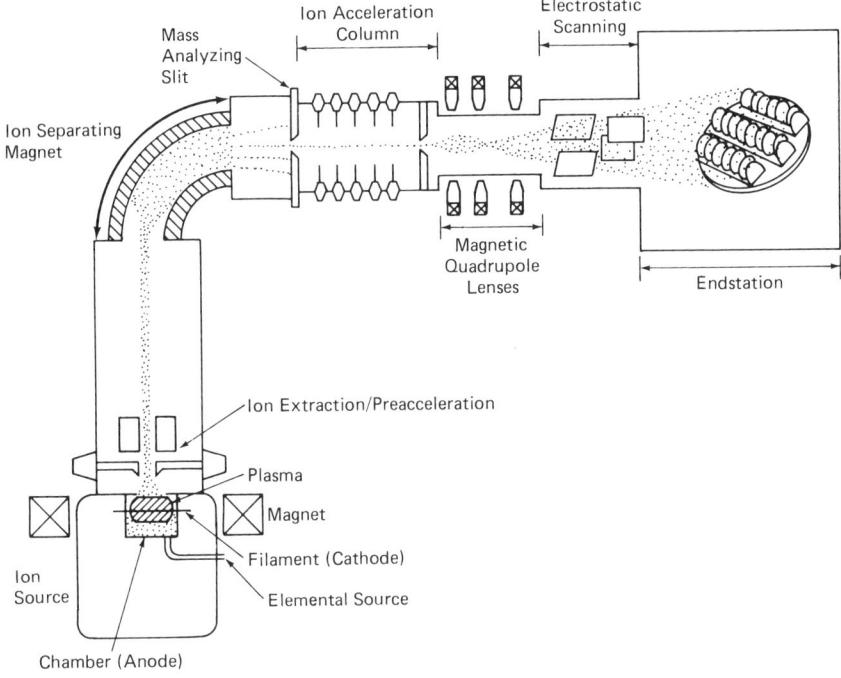

Figure 2-9. The production of energetic ions.[34] Courtesy of Sioshansi and *Materials Engineering*.[34]

target nuclei. Radiation damage at the entry surface is usually minimized by subsequent annealing at higher temperatures for control of diffusion of implanted ions. Laser annealing has been effective.

New surface metallic alloys and modified films have been achieved with a variety of super-hard surfaces, low friction bearings, wear resistant regions, and corrosion resistant alloys. Titanium alloy hip-joint prostheses have benefited from nitrogen ion implantation on their surface. In an environment of body fluids, the motion of the metal ball joint in the polyethylene socket can degrade the joint. Implantation of nitrogen ions greatly reduces the wear of titanium when it is in contact with polyethylene.[31]

There are other techniques of blending. The creation of ingenuous processing techniques appears to be keeping up with the diverse types of materials comprising metal/polymer composites.

38 METAL/POLYMER COMPOSITES

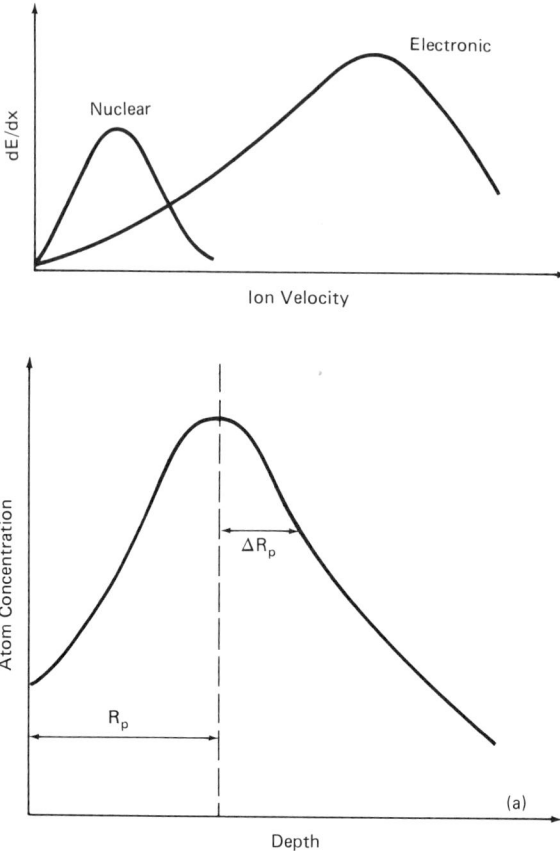

Figure 2-10. Ion implantation. Adapted from Reference 29. In top figure, as the ion enters the solid it is slowed down by electronic and nuclear stopping; at low velocities, nuclear stopping becomes predominant. In bottom figure, typical theoretical depth distribution of implanted atoms shows an average range of R_p and halfwidth $\triangle R_p$.

REFERENCES

1. Pease, L., *International J. of Metallurgy*, 22(3): 177, (1986).
2. Delmonte, J., *Metal Filled Plastics*, New York: Reinhold Publishing Co. (1961).
3. Hausner, H., *Handbook of Powder Metallurgy*, New York: Plenum Press, (1973).
4. Kamal, I. *Metals Handbook*, Vol. 7, 9th Edition, p. 606, American Society of Metals, (1984).
5. Bhattacharya, S., *Metal-Filled Polymers*, New York: M. Dekker, (1986).
6. Poster, A. R., *Handbook of Metal Powders*, New York: Reinhold Publishing Co., (1966).
7. ASTM-B-310, *Sintered Carbon-Steel Structural Parts* (1985).

8. Stern, A. *"Crushing and Grinding Practices"* Chemical Engineering, New York: McGraw Hill, (December 10, 1962).
9. General Elec. Co., G.B. 2,155,048A and G.B. 2,155,049A (February 5, 1985).
10. Edwards, J. and Wray, R., *Aluminum Paint and Powder,* New York: Reinhold Publishing Co., (1955).
11. Hildeman, G., et al. (Aluminum Co. of America) U.S. 4,435,213, (March 6, 1984).
12. *Japan-U.S. Seminar on Super Alloys,* Nagoya 457 Japan, (December 1984).
13. Rai, G., et al., *Journal of Metals* 37(b): 22, (August 1985).
14. Kubel, E., *Materials Engineering* 102: 25, (January 1985).
15. Hitachi, Japan patent 59,10,342 (January 18, 1984).
16. CA 100,143,009D, *Poroshk-Metall.* Kiev 184: 6, (1983).
17. Nippon Mining Co., Japan patent 53,217,608, (December 17, 1983).
18. Mitsui Mining, CA 100,160,707X JP 58,224,103, (December 28, 1983).
19. Kohrra, S. and Muto, M., *Conference at N.A.S.A.* (Virginia), (June, 1983); *"Recent Advances in Composites"* and A.S.T.M. STP 864, (1985); and Divecha, A., 4,060,412, (November 29, 1977).
20. Mayama, T. and Wase, H., to Toshiba Chemical U.S. 4,530,779 (July 23, 1985).
21. Olin Corporation, G.B. 2,152,533A, (May 10, 1984).
22. Lukon, et al., American Cyanamid Corp. European patent 248,384 (December 9, 1987), U.S. Application 869,518, (June 2, 1986).
23. Kenishizoko Photo Ind., Japan patent 61,72,587, (April 14, 1986).
24. Cameron, C. *Non-Glassy Inorganic Fibers,* NASA SP 5055, (August 1966).
25. Sagamore Research Conference, Syracuse University MET 661-601, (August 1959).
26. Milewski, J. and Shyne, J., SPI Reinforced, 20th Annual Section 6-C, (1964).
27. Mitsubishi Rayon Co., Japan patent 61,66,775, (April 5, 1986).
28. Bekaert, N. V., U.S. 4,664,971, (May 12, 1987).
29. Sealy, B., *"Ion Implantation Doping of Semi-Conductors,"* International Materials Review, 33 (1): 38, (1988).
30. Schoch, F. and Bartko, J., *"PPS Ion Implantation,"* ANTEC (SPE), 44th Annual, p. 273, (April 1986).
31. Picraux, P. and Pecky, P., *Scientific American,* 252: 102, (March 1985).
32. Todd, D., *Plastics Compounding,* 11: 33, (Jan/Feb 1988).
33. Anon., *Plastics Compounding Redbook,* 11(4) p. 94, Edgell Publications, Denver (1988/1989).
34. Sioshansi, P., *Materials Engineering* 104: 19, (March, 1986).
35. Seymour, R. B., *Additives for Plastics,* Vol. 1 and 2, New York: Academic Press, (1978).
36. Katz, M. H. and Milewski, J. V., *Book of Fillers and Reinforcements for Plastics,* New York: Van Nostrand Reinhold Co., (1978 and 1987).
37. Buck, M. and Suplinskas, *Engineered Materials Handbook,* Vol. 1, p. 851, ASM International, (1987).

3
MOLDING AND CASTING OF METAL/POLYMER COMPOSITES

Metals in particulate forms are blended with liquid polymers or solid particles of polymer, as described in the previous chapter. Applications of these composites as molding, casting, or adhesive compounds are numerous in the electrical and electronic industries and as maintenance materials. In the majority of these examples, the principal matrix is the polymer which provides continuity throughout the structure. This suggests that qualities such as adhesion, chemical resistance, coloration, reduced heat transfer, and electrical insulation are the primary objectives. The presence of metal particulates contributes to higher densities, improved heat stability within the thermal limits of the accompanying polymer, barriers to high frequency radiation, and the availability of magnetic properties. The contributions of metal particulates become more formidable when they are introduced as chopped fibers or pellets and hence make possible higher moduli of elasticity and better impact strength than is possible with unreinforced polymers. Impressive structural advantages accrue when sheets, rods, and continuous metal fibers, and metal oxide, or metal carbide fibers are part of the composite structure. The structural aspects of these combinations are cited in later chapters.

Continuing developments of particulate metal/polymer composites are introduced in this chapter to direct attention to new opportunities. These examples are selected from qualified sources, and focus upon the technical attributes of finely divided particles of polymer and of metals. For example, injection molding of metal powders with a small amount of organic binder yields products which, after high temperature sintering, possess a continuum of metal. This contrasts basically with the molded metal filled polymers characterized by a predominant polymer matrix. New high temperature resistant thermoplastic polymers described briefly in the previous chapter, demonstrate glass-transition temperatures up to 300°C. They will extend the temperature capabilities of organic polymers. Heat resistant phenolic resin molding materials will outperform most of the thermoplastics when placed under sustained physical stress at high temperatures. This is to be expected from the highly cross-linked thermosetting polymer. In processing by thermal blending, advantages lie with thermoplastics over thermosets, because they will not precure to an infusible state during processing.

Polymer structures may be composed of crystalline segments as well as amorphous segments. This will influence processing measures. These developments are significant for the growth of particulate metal/polymer composites. As the temperature gaps between metal processing and polymer molding are narrowed, there are more opportunities for physical commingling of materials.

Commercial metal filled molding compounds have established an important base in magnetic materials, and their characteristics are examined in Chapter 8. Liquid and paste-like composites of thermosetting epoxies are found in many commercial applications, and their applications are reviewed in this chapter.

Depending on the manufacturing process used (chemical, atomization, and grinding options were described in the previous chapter), metal particles are spherical, fragmented dendritic structures, platelets, or foraminous with dimensions from 1 millimeter down to 100 nanometers. Most metallic particles (excluding gold and silver) will have oxidized surfaces, traces of impurities from grinding, traces of lubricants added to minimize dusting, and certainly traces of moisture. Premium prices are paid to keep these contaminants at a minimum or within specific physical or chemical limits.

Polymers have their own classes of impurities which may or may not influence the blending operations. The range of molecular weights for a commercial polymer implies there are volatile low molecular weight fragments and more viscous high molecular components. Polymerization catalysts for some classes of thermoplastics may involve trace amounts of metallo-organic complexes. For the larger size applications of metal/polymer, most of the trace impurities present may be disregarded, but for the submicrometric areas of electronic units, the influences of trace impurities from either the metals or polymers must be considered.

PRESSURE MOLDING OF PLASTICS PARTICLES

The metal powder industries and the polymer molding companies are investigating the advantages and disadvantages of their respective processes. The plastics molding industry has historically used the following techniques:

1. Preheated and preweighed bulk thermoplastics such as bitumen and coaltar compounds are placed in cold molds and compacted in hydraulic presses. This process, referred to as *cold molding,* has been used extensively for molding storage battery cases. More recently, these cases have been formed from injection molded, chemically resistant polyethylene.

2. Granules, pellets or preforms of thermosetting plastics (including mineral, metallic, or organic fillers) are cured under heat and pressure in closed

steel molds, forming a substantially insoluble and infusible product. This process is known as *compression molding*. For further information upon plastics molding processes and equipment, and the many fillers and additives, see References 1 and 2.

3. When thermosetting polymer molding compounds of phenolics or epoxies, for example, are placed in heated cylinders and forced into heated dies, the process is called *transfer molding*. Two alternatives are shown in Figure 3-1, one with the molding pressure clamping the dies together and the other with a separate hydraulic plunger for transfer of molding materials. Pressures of the order of 1000 psi to 5000 psi (7 to 35 MPa) are applied. Temperatures vary considerably depending upon the polymer involved.

4. Pellets or granules of thermoplastic molding compounds are placed into hoppers and screw-fed or gravity-fed into a rotating screw within a heated cylinder and subject to one of the following alternatives:

- Heated thermoplastic in a molten state is forced through a heated nozzle into a closed steel mold (under clamping pressure). This is known as *injection molding*. To facilitate flow, heated runners may be used at the point of entry into the mold.
- Thermoplastic material is subject to high pressures and forced through a smaller extrusion die establishing the profile of the extruded material which is cooled rapidly. This is the basis of *extrusion molding*.
- A further alternative is the production of fine thermoplastic films (25 to 100 μm). Polymers such as polyethylene and polypropylene are melted and screw injected into dies, which will permit internal air pressure to expand the thermoplastic into thin films by a process known as *blow molding* (see schematic details in Figure 3-2). Improved physical properties may be expected from the thermoplastic films, which are stretched circumferentially and longitudinally during the blowing process. There is a large market for vacuum metallized thermoplastic film. This subject is treated in a later chapter on metal/polymer composites for shielding and packaging.
- Low pressure consolidation of finely divided, low temperature melting polymer powders should also be noted in this listing. This technique is used for powder coatings and in rotational molding (the activities of the Association of Rotational Molders should be noted here). Rotation of the mold allows uniform distribution of polymer particles. The successful production of large containers is facilitated, with the replication of exterior surface details only. Heated cast aluminum molds are favored in production.

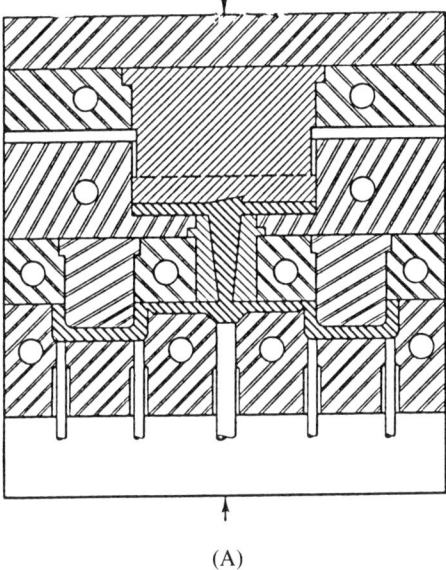

(A)

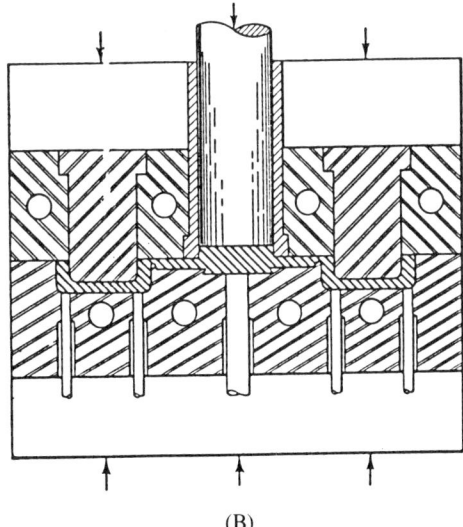

(B)

Figure 3-1. Transfer molding dies for plastics.[2]

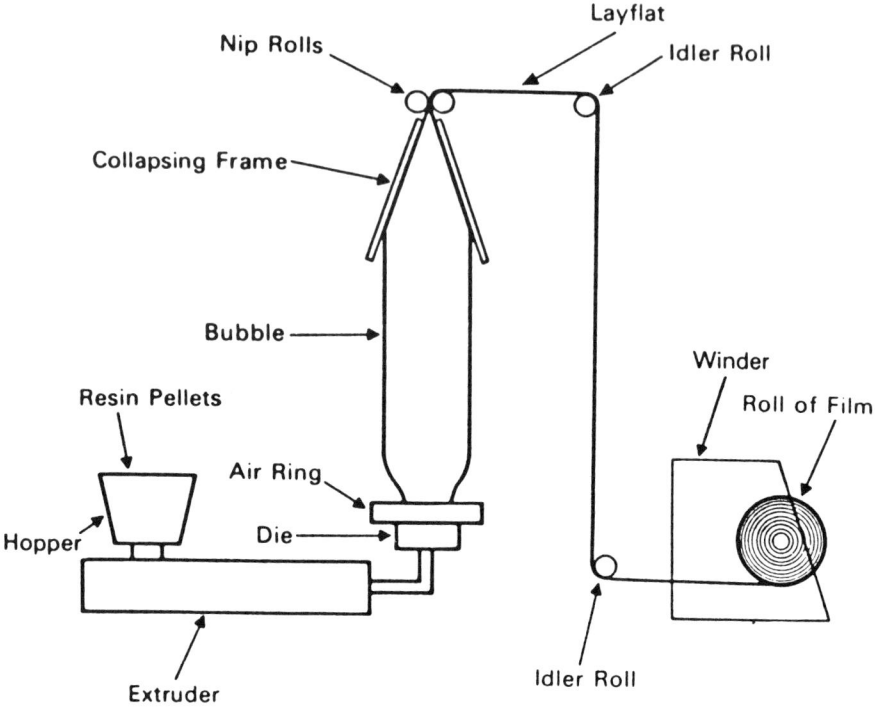

Figure 3-2. Basic elements of blown thermoplastic films.

PRESSURE MOLDING OF METAL PARTICLES

Metal powders and granules may be consolidated using the techniques described below. The metal particles are generally mixed with a small amount of lubricant before compaction.

1. Compaction of metal powders is usually conducted at hydrostatic pressures in the range of 10 to 60 tons per square inch (140 to 830 MPa). Shaped, low strength compacts are produced with the size and shape of the finished part, when they are ejected from the die. The compacts have sufficient "green" strength for handling and transport to a sintering furnace. They are brought up to sintering temperatures for the metal involved and organic material is volatilized. Figure 3-3 illustrates a typical compacting cycle for metal powders, as depicted by the Metal Powder Industries in their design bulletin.[2A] Compacting dies for shaping and compacting metal powders are usually fabricated from hard nitrided steel.

MOLDING AND CASTING OF METAL/POLYMER COMPOSITES   45

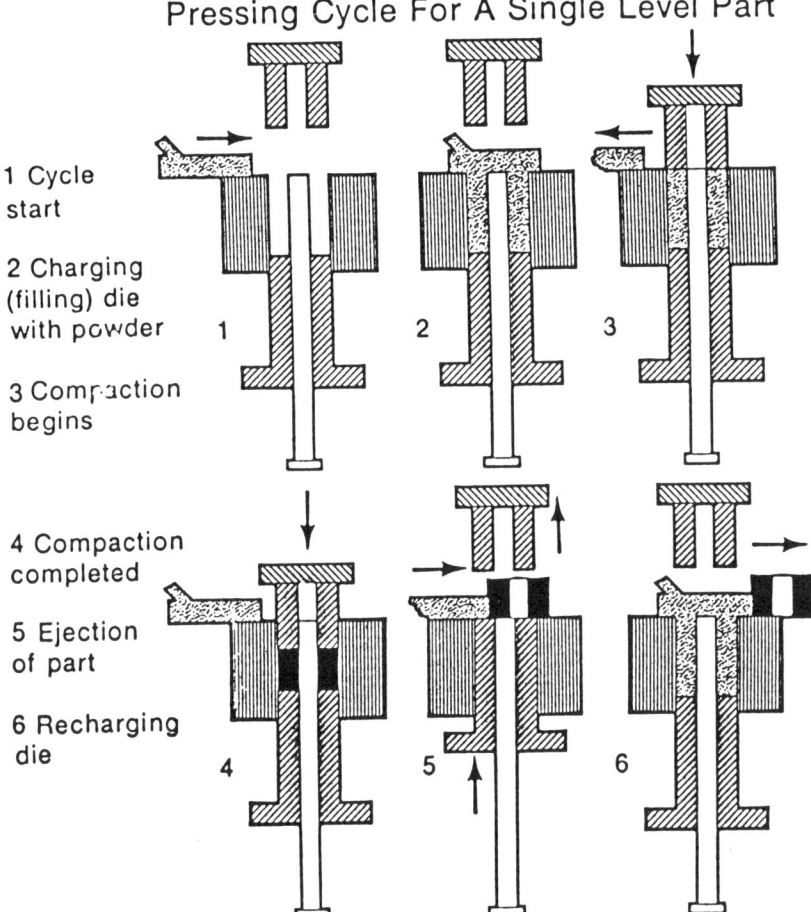

Figure 3-3. Compacting of metal powders. Courtesy of Metal Powders Industry Federation.[2A]

2. Isostatic pressing is performed in a pressurized fluid such as water or oil, which transmits pressure through a flexible membrane to the powder mass. Powdered metals are produced to near net sizes and shapes, including internal cavities. Metal powders with spherical or rounded particles are not cold compacted as the "green" strength of compacts is not adequate. Sintering of parts takes place in a controlled atmosphere furnace.[2A]

3. Hot Isostatic Pressing (H.I.P.) is performed in a pressure vessel contain-

ing a heated inert gas (argon or helium). Under uniform pressure of the gaseous medium and high temperatures, larger size compacts (up to one meter) may be formed. The compacts are formed at pressures up to 15,000 psi (103 MPa) and then exposed to sintering temperatures up to 2300°F (1260°C) for steel products. Densification achieved by H.I.P. is close to theoretical density for the metal. Good mechanical properties are attained.

4. Explosion compaction of metal powder compact is one of the new procedures under investigation. Because pressures released by explosives are high, theoretical densities of the metals are achievable. For Al-Si-9310 alloys, strength properties are higher than obtained in the wrought alloy.[6]

5. Much in common is shared by the techniques of zinc metal die casting and the injection molding of thermoplastics. The basic principle is the same—transfer of molten or semi-fluid material under pressure into a closed die which imparts dimensional details on the parts being produced. Die clearances, heating and cooling procedures differ significantly from plastic processing. In general, the chemistry of the organic plastic and its conversion to a solid body is more chemically complex. There are processing details used by the plastics industry which are transferable to the metal die casting industry and vice versa. Design, performance, and marketing objectives influence material selection.

Semi-solid thixotropic magnesium alloys have been injection molded at 580°C as reported by Dow Chemical Company.[5] This temperature is below the normal melting point of the alloy. The mold is filled in about 20 milliseconds at pressures up to 105 MPa (15,000 p.s.i.). Porosity of the magnesium molding is noted at 1.5% (about one-half of the usual die-casting porosity). Figure 3-4 shows a schematic view of the apparatus. Note the use of argon gas to establish an inert atmosphere, essential to high temperature processing of powdered magnesium.

Design recommendations for metal powder compacts and parts are illustrated in Figure 3-5. These self-explanatory details emphasize the draft or taper that must be present in the forming molds to permit ease of removal of parts from the tools. The green or unsintered strength of metal powder compact is not high until the welding of particles takes place at high temperatures below the melting range of the metal alloy. It is quite likely that the presence of a small amount of polymer powder will contribute to adhesive bonding and hence higher green strength of metal compacts before sintering. Figure 3-5 emphasizes the importance of avoiding undercuts which inhibit removal of metal compacts from the forming tool, and the desirability of maintaining uniform material thicknesses and avoiding sharp corners and abrupt changes in dimensions.[10] These design recommendations are similar to those used for the high temperature molding of plastics materials except that the lower pressure required for molding polymers permits more complex mold details.

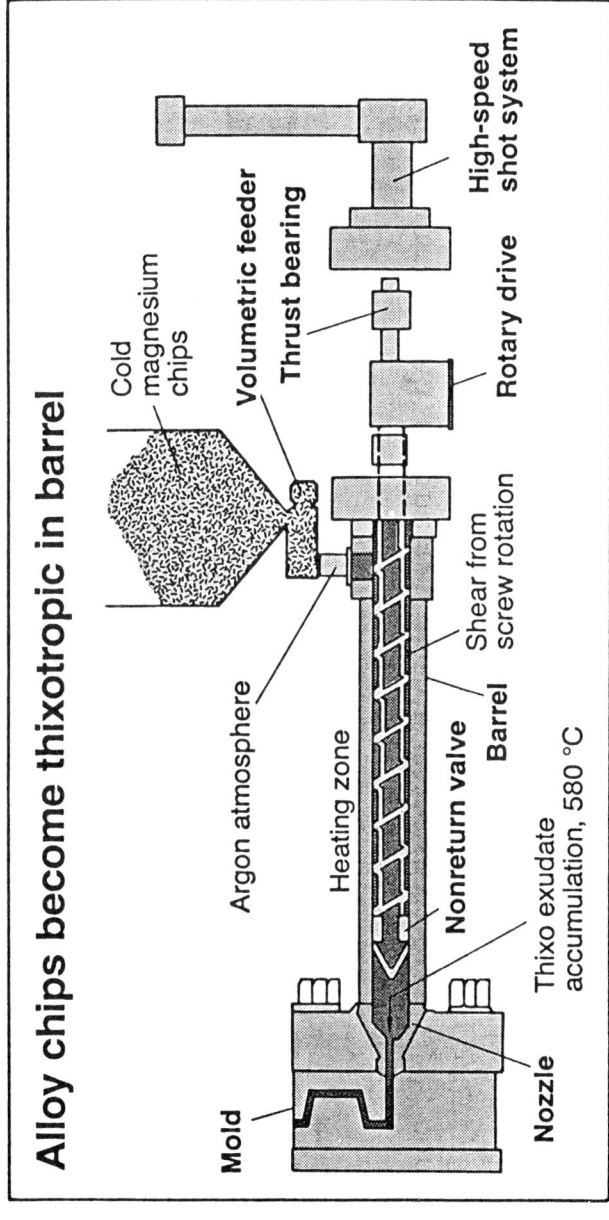

Figure 3-4. Injection molding of magnesium. Dow Chemical Co.[5]

48 METAL/POLYMER COMPOSITES

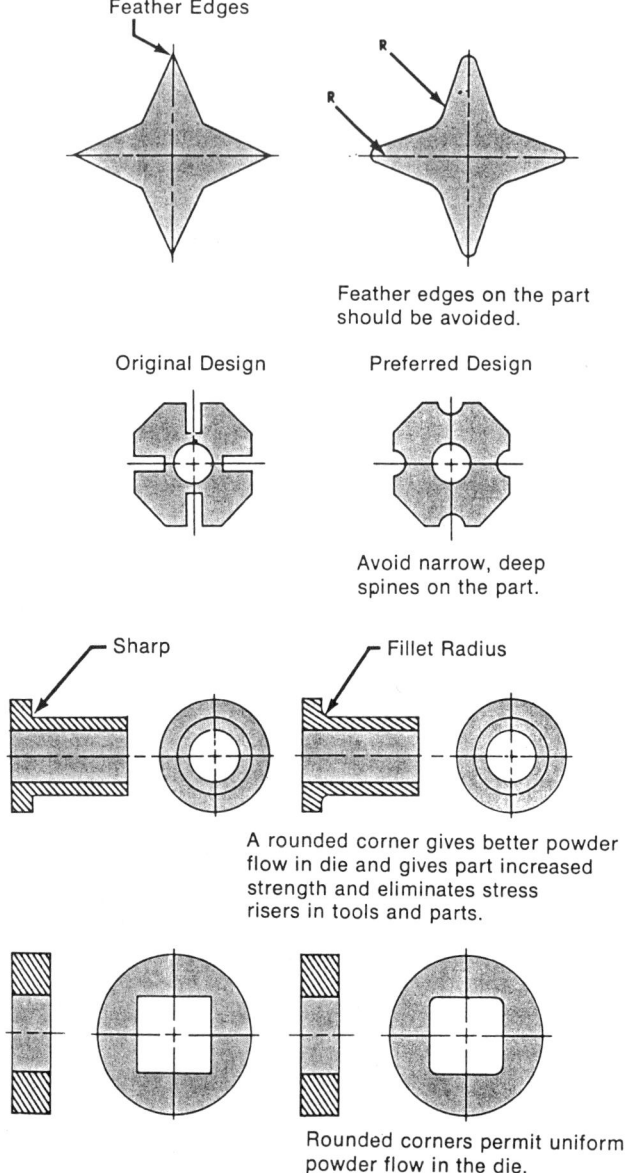

Figure 3-5. Design recommendations for molded metal parts. Courtesy of Hoeganaes Corp.[10]

MOLDING OF COMBINED METAL AND POLYMER PARTICLES

The processing of metal/polymer composites has proceeded in several new directions. The first one described below, injection molding of steel, requires the use of sintering temperatures to complete consolidation of the mass, because the metal matrix is dominant.

1. Injection molding takes place on fine metal powders (preferably under 10 micrometers) which are coated with a minimum amount of thermoplastic resin to bind the particles into a uniform mass. Some elevated temperature during processing will enhance the effectiveness of the organic binder. Parts are compacted close to net dimension, and consolidation is completed as the material passes through a high temperature sintering oven. A typical four-inch diameter steel gear produced in this manner is illustrated in Figure 3-6. Weight loss of initial material, by TGA tests, was approximately 9.5%. Linear shrinkage depends on the organic binder of the initial material. Up to 95% of theoretical steel density can be achieved.

The injection molding of metal powders with small amounts of organic binders to enhance the green strength raises the problem of removal of the organic polymer binder before high temperature sintering takes place. (For steel, hydrogen sintering for 1 hour takes place at 2050°F (1120°C) or 2 to 4 hours at

Figure 3-6. Injection molded steel gear.

50 METAL/POLYMER COMPOSITES

2500°F (1370°C) under vacuum.) In the Rivers process, a water soluble organic binder is used (methyl cellulose). As the water evaporates, there is an appreciable porosity through which the decomposition products can escape.[11] This reference draws attention to current practices, and data on injection molding of metal alloys. In another paper, feed stocks of an alloy of Fe-Ni-Mo were described as comprising 92% of the molding composition.[13]

2. There are many examples of powdered metals or conductive fibers which are added to moldable polymers for the purposes of providing shielding against electromagnetic radiation. This is an important subject and is discussed in Chapter 7. Injection and compression molding grades of plastics are used. A few examples are noted.

- Metals and alloys with melting temperatures under 400°C are molded with thermoplastics. Typical low temperature alloys are shown in Table 3-1. Powdered lead (9 kg) and 1 kg of polystyrene were mixed and molded, in one example, at 210°C.[7] The selection of a high density metal such as lead, would be indicated for a shielding requirement against x-rays. Lead filled rubber-like aprons furnish striking examples.
- Synthetic resins are injection molded with a powdered low-metal/polymer alloy consisting of 48% bismuth, 28.5% lead, 9.0% antimony, and 14.5% tin. Electromagnetic wave shielding is provided by the molded cabinets.[8]
- Molded parts which are used for EMI shielding are produced from polypropylene (65 pbw) and a tin-zinc alloy (35 pbw) (M.P. 210°C). These are blended at 250°C. Molded parts have a resistivity of 0.3 ohm-cm, a tensile strength of 4900 psi and elongation over 800%.[9]

Table 3-1 Melting Points of Mixtures of Metals*

ALLOY	
Lead—60% tin	190°C
70% tin	185°C
80% tin	200°C
Lead—50% bismuth	145°C
60% bismuth	126°C
70% bismuth	168°C
Lead—10% antimony	250°C
20% antimony	275°C
Bi-50; Pb-25; Sn-12.5, Cd-12.5	70°C
Bi-50; Pb-28; Sn-22.0	109°C
Sn-63; Bi-37	183°C
Sn-60; Zn-40	210°C

*Handbook of Chemistry and Physics, 67th ed. Section F-19. CRC Press 1987.

3. Shaped metal inserts of low melting temperature and small metal castings are being used as inserts to produce cavities within molded thermosetting plastics. Upon completion of the molding operation, the metal alloys are melted out, and recovered for reuse. Fully enclosed cavities for small pumps are produced in molded plastics, without the expensive tools required by more conventional techniques.

4. A novel method for processing high molecular weight polyethylene, polyimides, polyamide-imides, and polyphenylene sulfides which possess high viscosity during molding at high temperatures, introduces powdered iron (10% volume loading) into the composition. The presence of the iron enhances heat transfer, and generates heat losses in an alternating magnetic field. Cold compaction under a pressure of 220 MPa is followed by heating the polymer/metal mixes at about 175°C for several hours.[3] Microwave polymerization of epoxy-iron composites at 2.45 GHz inside a wave guide, has also been reported.[4] As expected, increasing iron particle content shows a marked increase in rate of polymerization, earlier realization of high temperatures, and better molding.

The application of high frequency radiation to cure polymer adhesives has been practiced since the late 1940s. At that time there were advantages to bonding wood structures with adhesives that become warm rapidly under high frequency radiation. This was particularly important as a source of heat for adhesive assemblies difficult to reach by other means. Much of the R.F. heating was done at 13.6 and 27.32 megacycles. More recently, the advantages of adding powdered iron to thermoplastic adhesives has allowed the use of lower frequencies in the 1K to 10K hertz range. Efficient bonding of materials has become possible, particularly on irregular shaped adhesive assemblies. A typical system is the Ermabond process of Ashland Chemical Company where the exposure to high frequency radiation develops heat in the form of hysteresis loss in the metal particles contained in the thermoplastic polymer. This principle has also been applied to the heating of iron filled molding compounds described in the previous paragraph.

5. Another successful example of powdered metals and powdered binders, one that does not require molding pressure, is the field of xerography. Carlsen's early developments in 1938 were followed by association with Haloid Corp. (later to become Xerox Corp.). The first production model using the techniques appeared in 1959.[14]

The use of powdered metals and finely divided thermoplastic binders is of major importance to xerography and to developments in copying machines. Finely divided iron powder mixed with thermoplastic binders, when selectively applied to paper stocks, fuses under concentrated heat and bonds to the paper surface. When an optical system is coordinated with electrostatic attraction, a typed page may be reproduced with fidelity. The Canon copier contains a "charged couple device array" in which a computer chip converts light into digital information. This data is transmitted to a laser printer. Data can be stored if desired with the latest development and copies made without the originals. The market for copying or reproduction machines continues to expand rapidly, particularly with laser printing. Powdered iron and thermoplastic resin binders are the key to successful processes and represent successful dry blends of metal powders and polymers.

SIMILARITIES OF PROCESSING METAL AND PLASTIC POWDERS

In reviewing the basic techniques of metal powder injection molding (MIM) and injection molding of plastics we must keep in mind that there are many parallel paths that are being followed. This is particularly apparent in the development of processing equipment and in the evidence that there is a sharing of manufacturing experiences by polymer and metal powder companies. Roll of the Metal Powders Industries Federation acknowledges that the separation between metals and nonmetals is becoming "fuzzier." He notes, "We will share closer relationships with plastics and a host of composite ceramic structures," and cites as an example the fact that metal injection molding is combining the best features of ceramics, powder metallurgy and plastic technology.[12]

The addition of small amounts of organic binders to ferrous powdered mixes has produced some interesting results. In one such example the ingredients of a regular powder metal mix were supplemented by a small amount of binder. Mixing took place in 500 lb. batches in a double cone blender. Studies have shown reproducible results for all of the common alloy ingredients typically used in iron powder mixtures.[12] The basis of improvement is thought to be a combination of the binder effect in agglomerating the fines, and the fact that the binder is a tack-free solid at ambient temperatures. The original objective of the research which led to the binder application was to develop premixes which would be more economical to process than existing metal premixes.

BASIC CONTRIBUTIONS OF METALS AND PLASTICS TO MOLDED COMPOSITES

Powdered metals and finely divided polymers are pre-blended into composites by extrusion blending, for example, and parts are shaped in molds under heat and pressure to cure and set the polymer matrix. Electromagnetic devices offer many opportunities for ferrous metals and polymers. Developments of resin bonded magnetic materials fulfill important commercial roles. The metal component may be expected to alter the original attributes of the composite in one or more of several ways:

- The molded part requires the contribution of the heavier metal to a higher density part or component (as in a flywheel, for example).
- The presence of metal contributes to the attenuation of electromagnetic radiation. Effective shielding procedures are discussed in Chapter 7.
- The presence of metal enhances heat conductivity of the molded part, providing a heat sink to reduce the temperature rise of components in contact with electrical conductors.
- The presence of conductive metal particles contributes to greater electrical loss factors in the molding material, an asset when this characteristic is desired (in high frequency heating).
- Reduction of electrical volume resistivity and surface resistivity by metal powder additions is a useful procedure with molding compounds when static charges are to be minimized, though generally at some sacrifice of high electrical resistivity of the polymer.
- Radiation from nuclear sources can be reduced. Material selection may require the use of elements such as boron and cadmium with high neutron capture cross-section, or when gamma radiation is the exposure, a high density metal such as lead.
- Finely divided spent uranium-238, which has a density 80% greater than lead, has been used between steel plates for projectile shields on U.S. Army tanks. The residual radiation is very low and not harmful to personnel.

However, metal powders are not usually blended into polymer molding materials for added strength considerations, such as tensile strength or flexural strength.

The contributions of plastics to metals may be summarized as follows:

- The presence of small amounts of specific polymers will enhance the integrity and mechanical qualities of pressed metal compacts before they are fired at high temperatures.
- Abrasion and weathering resistance offered by plastic coatings are invaluable to thin vacuum deposited metal films on organic and inorganic substrates.
- Magnetic particles suspended in polymer binders offer more versatile products for magnetic devices and magnetic tapes.
- Controlled porosity may be obtained in metal powders through the use of specific polymer particles, which may be eliminated during sintering.
- Electrical insulative protection is offered to many metal electromechanical devices.
- Decorative qualities available in polymer/metal composites are virtually unlimited.

54 METAL/POLYMER COMPOSITES

- Thermal insulative qualities of most polymers enhance the serviceability of metal products.
- Polymer adhesives contribute to the assembly of metal components. Automotive, aerospace, mining, and commercial markets require resilient adhesives to absorb shock and vibration.
- Plastics processing techniques are more amenable than metal processing to the inclusion of fiber reinforcements for added strength and stiffness.

ENCAPSULATION OF METAL COMPONENTS BY PLASTIC MATERIALS

There are compelling reasons, as noted above, for the preparation of composites of finely divided metal alloys and powdered polymer compositions which are processed either by molding at high temperatures, or by casting composites of metal powder in liquid polymers. In addition, there are a large number of fabricated electrical and electronic units which are encapsulated by polymer mixtures and given their form or shape in a mold. Molding compounds of polymer powder mixtures are used extensively for encapsulation to form metal/polymer electrical composites, as well as liquid systems of polymers. For many components, the molding compounds are preferred, as they have been adjusted for rapid curing at high temperature and minimum shrinkage. The appearance of the final composite assembly, as shown in Figure 3-7, has sales appeal because it hides the wire coils, terminals, and insulation used to prepare the assemblies for encapsulation. Many thousands of small transformers, capacitors, relays, magnetic devices, circuits, inductors, etc. continue to be encapsulated in this manner. Dip coating and powder coating for encapsulation of electrical components are discussed briefly in Chapter 5, though this form of encapsulation involves a random coating, not a precise shape as defined by a mold.

While the role of the metal component is not always as significant as its role as an electrical component, it may fulfill important mechanical functions as an insert. Some of these encapsulated inserts will be discussed in this chapter, as they too are part of the metal/polymer composite profile. Furthermore, there are increasing numbers of examples of low melting point metal alloys which are removed by being melted out from the completed molded part.

The encapsulation of electrical components and fabricated assemblies of metals, ceramics, and insulative substances are performed principally by the molding of epoxy compounds. Phenolic molding compounds also have good performance characteristics particularly at high temperatures up to 150°C. Epoxies and phenolics are essentially room temperature stable-one component systems, with built-in curing agents, and are adaptable to the activities of

MOLDING AND CASTING OF METAL/POLYMER COMPOSITES 55

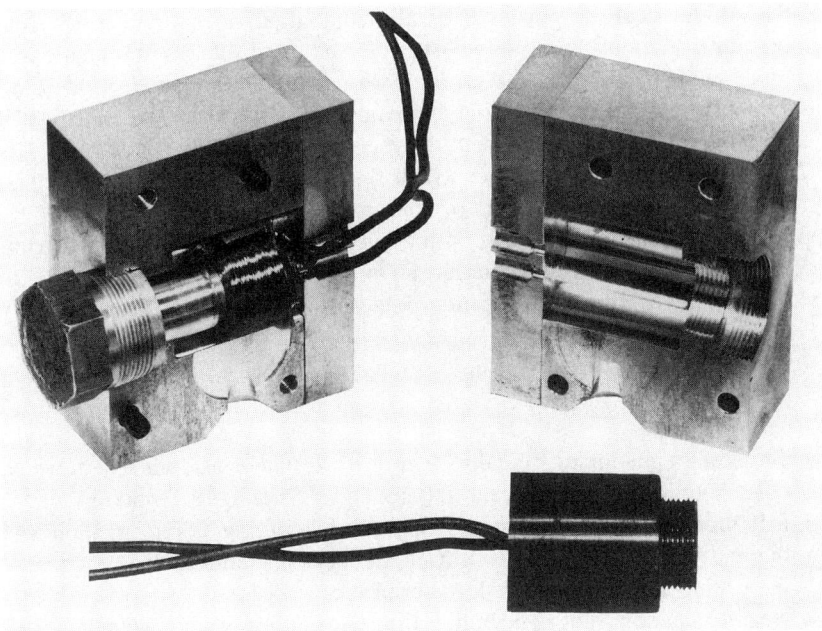

Figure 3-7. Steel mold and transfer molded EPOCAST "400" material. Courtesy of Furane, Division of Ciba Geigy.

plastic molding shops. However, the better adhesion and lower cure shrinkage of epoxy resin systems have propelled these materials to the forefront of encapsulation molding of electrical units. In recent years, the epoxidized phenolic "novolacs" have been the material of choice. Their improved heat resistance over earlier epoxy compounds accounts for this. New heat resisting thermoplastic polymers such as polyetheretherketone, polyphenylene sulfides, polyimides, polysulfones, poly (p-phenylene benzobisthiazole) are popular as shown by their performance in aerospace structures as laminated composites and adhesives. Their usefulness as encapsulants is being established in both the U.S. and Japan. The Japanese are aggressively identifying problem areas associated with encapsulant materials and are allocating the necessary resources to effect solutions.

Liquid encapsulants for electrical and electronic devices have utilized epoxy resins, polyester resins, and silicones. Rubber-like silicone resin encapsulants will perform satisfactorily at low and high temperatures, and withstand thermal cycling better than other more rigid materials.

Before the introduction of molding powders of epoxy resins for the encapsulation of electrical component assemblies, liquid polymer formulations were dispensed on assembly lines in one of two basic encapsulating formulations:

1. One-component systems with a curing agent (dicyandiamide, for example) which would not react at room temperature, but which would be activated when exposed to elevated temperature.

2. Two-component systems, in which the curing agent, generally a liquid polyamine derivative, is kept out of contact with the polymer until ready for application. When the resin and curing agent are mixed in the desired stoichiometric ratio, cure would proceed at room temperatures (typically 20° to 25°C) and the hardened insulation would form. Exothermic heat generally raised temperatures during cure.

Encapsulants are solvent-free and formulations prepared with the following objectives in mind:

- Good impregnation of wound wire coils to eliminate dead-air spaces which contribute to the development of "hot spots" among the windings.
- Minimum shrinkage. Most polymers shrink during polymerization and thermally contract as they cool from higher temperatures. The presence of inert mineral fillers helps reduce shrinkage.
- Heat dissipation, during operation from assemblies of wire wound devices will contribute to longer life and effectiveness of the unit. The selection of fillers is influenced by their heat capacity and thermal conductivity. The addition of fine powdered metals to the encapsulant generally establishes a good "heat-sink" and contributes to lower operating temperatures and better performance.
- Good adhesion is a basic requirement, though the diverse materials present in an electrical assembly makes this a difficult criterion to fulfill. Trace amounts of chemical coupling agents from silicone or titanate derivatives have contributed to improved adhesion to the assembly being encapsulated. Empirical studies show that a balance of different particle sizes will often yield a superior product.
- Complete tight sealing reflects good adhesion, good penetration, minimum shrinkage, and absence of micro-cracks at operating temperatures. The penetration of moisture or harmful agents is avoided through careful selection of encapsulants for electrical devices. Micro-cracking may be due to the low elongation of the polymer matrix. Small (less than 1 to 2%) additions of fine fibrous fillers have been known to alleviate this condition.

The above objectives, which have great significance in larger electrical devices, are also applicable to smaller micro-electric units embodied in printed circuit boards and semiconductor devices. The fact that such units are small, does not eliminate the factors of shrinkage, heat dissipation, and microcracking. These phenomena do take place on a microscopic scale and are contributory to device failure. Metal/polymer assemblies, particularly for electrical components must be produced with the objectives noted above. The objectives are also applicable to the semiconductor and electromagnetic devices discussed in later chapters.

Plastics packaging of electronic devices is generally accomplished by encapsulation in molded epoxy as illustrated in Figure 3-8. Either with the aid of metal pins, or threaded metal inserts, electrical contacts are established with the electrical components inside the encapsulant. These devices number in the thousands and are representative of metal/polymer composites. Detailed analyses are conducted upon the early models of such units. Thermogravemetric (TGA), thermomechanical (TMA), ion chromatography, scanning electron microscopy (SEM), and energy dispersive x-rays are among the procedures used to study the materials in the assembly.[16]

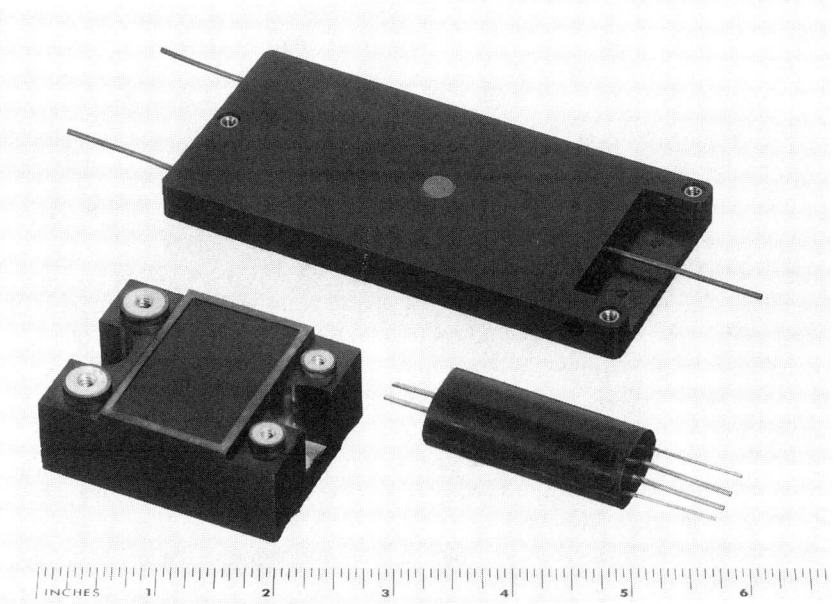

Figure 3-8. Electronic parts encapsulated in epoxy molding compounds.

METAL INSERTS

Metal inserts are encapsulated by polymers for a variety of purposes. The formation of molded containers with internal cavities may be accomplished by the blow molding of thermoplastics, or the use of temporary metal inserts, cast to define the shape of the cavity. These inserts are later melted out of the molded plastic part, which has formed about the metal insert. This latter technique calls for a metal alloy which has a melting temperature above the molding temperature range of the plastic material, but also a temperature low enough to fluidify the metal without adversely damaging the plastic molding material. Many low cost metal alloys of tin, lead, antimony, zinc, and bismuth are available for this purpose. This process is comparable to the "lost wax" process for the casting of metals. The liquid metal alloy is recaptured for further use. The table shown earlier in this chapter suggests candidates for low melting metal alloys.

Metal inserts fulfill many purposes in polymer structures as they bring to the composite qualities not necessarily possessed by the non-metallic material. This has been shown to be abundantly clear, particularly in reviews of electrical applications. There are applications which arise in the use of composites in which the presence of a metal component would provide properties not possessed by the polymer. These include by way of functional examples:

- Fine ribbons of conductive aluminum or copper, or even electrical high temperature resistive wire, have been inserted between strips of resin impregnated glass fiber tape. Other examples include thin flexible electrical heating elements, used as inserts, which have been prepared from fluorocarbon polymers combined with acetylene black (See Chapter 4). Coordinated with strips of conductive metal, they are designed to produce 10 to 60 watts of heat per square foot of product.[17] Metal flakes were blended with a polyamide-imide polymer in N-methyl pyrrolidone to form electrical resistance heating inserts in another example.[18] The subject of high electrical conductivity is treated in Chapters 4 and 7.
- Plastics jigs and fixtures are fabricated for precision drilling, welding, or dimensional check positions. Metal inserts at these positions provide areas less susceptible to wear and tear.
- Metal inserts are encapsulated during molding operations, particularly for attachment purposes. Molded plastic threads do not hold as tightly as metal threads due to their inherent creep and cold flow. Brass inserts molded in phenolic plastic 50 years ago show no evidence of deterioration at the interface. Metal inserts are usually grooved or knurled to furnish mechanical reinforcement on the interface. High frequency ultrasonic staking of inserts is also practiced on the thermoplastics.

- Metal rods are included in new designs of molded plastic tote boxes, providing stiffness and strength. Rods or structural shapes of brass, steel, or aluminum are placed as inserts in the molds. Steel wire reinforcements for laminated plastics have been known for sometime,[15] though they have been supplanted as inserts by high modulus graphite fibers.
- Rotational wear on plastics bearing surfaces may be avoided through judicial positioning of metal bearing inserts within the molded part.
- Kirkwood has perfected molded commutators for high performance power tools and appliance applications. Copper bars separated by mica strips are mechanically held with glass reinforced molding compounds. An assembled, integrated commutator is illustrated in Figure 3-9.

The assembly of metal components into plastic structures has been followed extensively since the early 1940s. This practice has been used with many molded plastic polymers to meet technical and esthetics requirements of the burgeoning plastics molding industry. The reason for the use of metal inserts is very clear—to provide reinforcements at points of high physical stress. A threaded attachment machined of brass or steel provides a more secure anchor which is not subject to the creep or flow that accompanies a threaded feature molded from a plastic material. Metal inserts are common in thermosetting phenol formaldehyde, and urea, or melamine formaldehyde components, as

Figure 3-9. Molded commutator for electric motors, comprising copper strips assembled with thermosetting molding compounds. Courtesy of Kirkwood Commutators, Ohio.

well as in injection molded thermoplastics. These inserts replace non-metallic threads molded in situ. The metal inserts are usually positioned on pins during molding.

The use of metal inserts as points of attachment in polymer structures also is practiced in the design of epoxy laminated tooling. Although large laminated tooling structures are produced in limited quantities, the need for metal drill bushings and metal attachments still exists and the metal components are "potted" into position, usually with the aid of a room temperature setting liquid or paste epoxy resin formulation. Literally hundreds of metal inserts may be used in such structures. Heat may arise due to frequent use of welding fixtures or in high speed drilling operations. If damage occurs to the metal insert, it is readily replaced and repotted into the cast or laminated plastic structure.

METAL POWDERS AND LIQUID POLYMERS

The addition of metal powders to liquid polymers is used for preparing castable or paste-like metal/polymer composites. While molding compounds may achieve a certain degree of fluidity at high temperatures and high pressures, casting techniques are generally associated with room temperatures (ca. 25°C) and they produce compounds which are pourable because of the lower viscosity. These composite products are widely used for maintenance purposes as is seen in the examples given. The polymers available for pouring are limited to low molecular weight thermosetting resins rather than thermoplastics. Thermoplastics are usually polymerized into solid high molecular weight materials, which are reduced to small particles and mixed with metallic powders, as described for molding materials. While a few liquid thermoplastic monomers such as styrene and methyl methacrylate, accompanied by curing agents (such as organic peroxides), may incorporate metallic fillers, they do not have the qualities and thermal stability which characterize cured thermosetting polymers. Thermoplastics in most instances may be blended with solvents, to which metal powders can be added, though the presence of volatiles in the polymer casting matrix is not desirable and must be removed completely.

Types of thermosetting liquid polymers which have been used with metal powders include the following materials:

A. Polymers which liberate water, or volatiles during cure, as a consequence of condensation reactions, include:

- Phenol-formaldehyde resins
- Urea-formaldehyde resins
- Melamine-formaldehyde resins
- Furfuryl alcohol systems
- Glyceryl-phthalate systems

B. Thermosetting polymers which are liquid and do not liberate volatiles during cure include:

- Epoxy resins such as diglycidyl ether of bisphenol A are available as fluid, low molecular weight liquids. They are frequently combined with reactive (non-volatile) diluents of glycidyl ethers, cycloaliphatic diepoxides, and curing agents such as nadic methyl anhydride or hexahydrophthalic anhydride to form various fluid casting compounds (usually less than 2,000 cps at 25°C).
- Polyester resins from polyglycols such as polypropylene glycol and isophthalic acid. For casting, polyesters are solubilized in monomers such as styrene or diallyl phthalate, and cured with organic peroxides plus accelerators.
- Polyurethane resin systems are prepared from methylene diisocyanate or toluene diisocyanate (T.D.I.) and polyglycols (including high molecular weight materials). Urethane modified epoxy coatings and adhesives for metal have been very successful. "Adiprene L" of DuPont prepared from T.D.I. and polyoxybutyleneglycol was one of the earliest urethanes used with epoxies.

There is a wealth of published literature on the polymers noted above and many opportunities for research into new products. Trends favor multipolymer blends, particularly among thermosetting resins, where for example, the pendant groups of a polymer may provide functional reactive positions which would be amenable with another polymer. Epoxy-urethane systems are adjusted to increase stiffness through the epoxy fraction or sacrifice rigidity for the toughness and elongation of the urethane component.

Type A thermosetting condensation type phenolic polymers were among the earliest polymers to have been used for casting. The catalysts to promote cure are influenced by the chemistry of the initial liquid phenolic resin. Strong acid catalysts are untenable for the phenolics if metal powders are to be added. On the other hand, the presence of strong alkaline catalysts render the phenolics unsuitable because of poor environment resistance of the cured material. More adaptable are formulations of urea and melamine formaldehyde, which in the presence of water soluble catalysts can tolerate metal powder additions. The fluidification is provided by the presence of water. In compounding, when anhydrous calcium sulfate is present as a filler, the excess water participates in the reaction to form hydrates $CaSO_4 \cdot 2H_2O$ which contribute to the physical strength of the structure and consume "free" water. Compounding the melamine formaldehyde with aluminum powder, porous inorganic spheres, and calcium sulfate hemi-hydrate forms the basis of a commercial product "Alumocast" which demonstrates the low density of this product, a cast metal/

62 METAL/POLYMER COMPOSITES

polymer composite. The flaky nature of some of the aluminum filler contributes to the water resistance.

Of more immediate interest for the casting of metal powders are the type B thermosetting polymers such as polyesters and epoxy resins. The advantages and disadvantages of several resin systems for the casting of composites of metals and polymers are shown in Table 3-2. There are many reactive diluents such as phenyl or benzyl glycidyl ether which can lower the viscosity of the epoxy resin systems, and in the case of polyesters the effective styrene monomer concentration can be adjusted to a desired viscosity. Typically 500 to 3000 centipoises before filler additions would be acceptable. The viscosity of phenolics and urea-formaldehyde cannot be conveniently decreased without the addition of solvents or plasticizers which would degrade the properties of the compound.

Finely divided inorganic fillers, in addition to metallic fillers, offer the advantage of reducing settling tendencies of the denser metallic fillers. If some settling takes place during long storage, a soft pack at the bottom of the container facilitates the redistribution of the components of the composite mix. Finely divided silica fillers such as "cabosil" or "bentonite" are used to aid the suspension of metallic particles. When castings of metal powder filled polymers are made, there is more separation and settling of the dense metallic particles when the viscosity of the pourable composite mixture is low. Better particle distributions may be expected from more viscous mastics.

Table 3-2. General Comparison of Cast Polymer/Metal Composites

	EPOXY RESIN SYSTEMS	POLYSTER RESIN SYSTEMS	POLYURETHANE RESIN SYSTEMS
Effect on metallic fillers	Minimal	Minimal	Some dependency on filler*
Exotherm when R.T. cure is practiced	High dependency on mass	High dependency on mass	Low to moderate
Shrinkage during cure	Low	High	Low
Hardness (Durometer)	Shore D over 90	Shore D over 90	Shore D adjustable from soft to hard by selection of raw materials.
Adhesion	Good	Fair	Excellent
Heat resistance	Good	Good	Poor
Environmental resistance	Good	Good	Good

*Some metallic ions are effective catalysts for curing polyurethane resins.

MOLDING AND CASTING OF METAL/POLYMER COMPOSITES   63

Among noteworthy applications of metal-filled castable epoxy composites are nonmetallic tools weighing several thousand pounds. These are used by aircraft manufacturers for the fabrication of sheet metal parts. The writer has prepared formulations from aluminum powders for lighter weight tools, and has prepared formulations from steel powder filled epoxy tools for high impact forming.[19, 20] The rationale supporting these choices is that the composite tool can be prepared and available for use in a fraction of the time required for a metal tool. A typical composite tool of epoxy resins and metal powders, weighing over 1,000 pounds is shown in Figure 3-10. With the large loading of inert fillers, the high exotherm of epoxy castings will be ameliorated. At the same time, adjustments can be made in the room temperature curing agents, and part of the reactive polyamines can be replaced by a tertiary amine such as triethanolamine, if necessary.

The steel rod in Figure 3-11 illustrates the durability of a metal filled epoxy grouting compound. It was placed in service over twenty-five years ago. The vertical member is a galvanized steel fence rod set in place by pouring in situ, an aluminum powder filled casting compound, with a room temperature setting polyamine curing agent. Epoxy resin systems have available a large number

Figure 3-10. Plastic drop hammer die, filled with Epocast 11D. Rohr Aircraft Corp.

64 METAL/POLYMER COMPOSITES

Figure 3-11. Galvanized steel bars set in epoxy mortar, excellent after 25 years.

of low temperature polyamides and polyamine adducts for R.T. curing. For an up-to-date reference on epoxy resins see the second edition of May's book on epoxy resins. (Ref. 3-21).

Commercial resin systems featuring miscellaneous metal powders dispersed in a liquid polymer have been offered under the trade names Epox-E-Cast and Devcon. The properties of Devcon materials appear in Table 3-3. The table shows the physical properties that can be obtained with several metal filled epoxies. The reference to "Plastic Steel 5-Minute" describes one of the compounds which is formulated to cure quickly and set rapidly in five minutes. Such materials are intended for small area, limited application. The mixing time of the two components will use some of the available application time.

Paste-like epoxy resin systems are usually prepared for application at room temperature, hence the necessity for a two-part system. When the curing agent is added, the polymerization commences and the paste or putty must be ap-

Table 3-3. Typical Physical Properties, Metal Filled Epoxy Compounds (7 days, room temperature)

	PLASTIC STEEL PUTTY	PLASTIC STEEL 5-MINUTE	ALUMINUM PUTTY	BRONZE PUTTY	STAINLESS STEEL PUTTY
Mix ratio, wt.	9.0:1	1.7:1	9.0:1	9.0:1	9.0:1
Resin:hardener vol.	2.5:1	1.0:1	4.0:1	3.0:1	3.0:1
Specific volume, in³/lb.	11.9	14.1	17.5	12.4	12.4
Viscosity w/ hardener, CPS	Putty	Putty	Putty	Putty	Putty
Pot life, minutes	45	5	60	35	45
Operating temperature, maximum °F	250	200	250	250	250
Compressive strength, psi ASTM D 695	8,260	10,400	8,420	8,540	8,900
Cured hardness, Shore D ASTM D 1706	85	86	85	85	85
Flexural strength, psi ASTM D 790	5,600	7,680	6,760	6,180	6,300
Adhesive tensile shear, psi ASTM D 1002	2,800	2,026	2,600	2,680	3,040
Cure shrinkage, in./in. ASTM D 2566	0.0006	0.0090	0.0008	0.0010	0.0010
Color	Dark grey	Dark grey	Aluminum	Bronze	Grey

Courtesy of Devcon Corp., Danvers, Mass.

plied. When rapid cure is required, high temperature cure agents are used. There are two major uses for the epoxy paste products: 1) Automobile body solders, and 2) Hull smoothing compounds for the ship building industry. The experience gained from these fields is applicable to a wide assortment of large metal parts fabricated from metal castings and sheet metal parts. For lower cost products there are paste-like materials from polyester resins and polyvinyl systems, the latter depending upon solvent evaporation to leave a layer of metal powder filled polyvinyl resin. Polyurethane resin pastes have exceptionally good impact qualities and are adaptable to assembly line production.

A valuable analysis of automotive requirements for body solders is contained in a paper by Peerman and Floyd.[22,23] Important requirements for automotive body solders may be listed as follows:

- Rapid cure of about 2 minutes at high temperatures on the automotive production line. Moving assemblies cannot wait for slow curing adhesives to set. Epoxy resins cured with high temperature cure agents will set rapidly at temperatures of 150°C.
- No sagging, blistering, foaming or cracking of the solder during cure must take place. This is not an easy requirement. Thixotropic pastes which may be non-flow at 24 to 27°C may flow at elevated temperatures.
- Cured pastes should have a high degree of impact strength under the conventional environmental temperatures (e.g., down to −30°C).
- Dinging or reverse peening of cured body solders to conform to surrounding contours necessitates exceptional adhesion and impact resistance.
- Under no conditions should steel panels which have been bonded or soldered react with the exterior coating applied over them, nor should any tendency to bleed through be evident.
- Uniformity in appearance and freedom from lumps and foreign particles is a foregone conclusion for a high quality body solder.

An examination of the many formulations for body solders discloses that metal-filled epoxy pastes offer those qualities most desirable for automotive practice. Localized heat from infrared lamps may be used to expedite the cure of the organic type solder. Lead based solders are 10 times more dense than the organic solders and actually more costly then the plastic solders on a volume basis.

To develop good body solders from metal pastes, the following measures are usually followed:

1. Resin component to hardener component (such as the polyamides or imidoazolines) are employed on an equal volume basis. Distinct color differences of resin and hardener phase facilitate mixing. The metal filled portion is silver or black in appearance with the hardener a distinctly lighter color.
2. After mixing, there must be good trowellability to aid spatula or gun application; furthermore, the ability to stick to vertical surfaces must be apparent during cure and at higher temperatures.
3. For optimum impact resistance and adhesion, metallic fillers such as powdered iron and its oxide are used, sometimes in combination with fibers (e.g., glass fibers have proven useful), polyaramid fibers are also used.

Besides the professional large volume automotive uses of body solders, amateurs can attain semiprofessional results. This is exemplified by sports car magazines which extoll the virtues of such adhesive pastes. Car enthusiasts concerned with repair or basic changes in body styling use epoxy steel pastes to complete their "customizing" operations. Metal powders and organic plastics are well known for their fortuitous union to form outstanding composites.

Hull-Smoothing Pastes (Metal/Polymer Composites). Hull-smoothing pastes used by the ship building industry are akin to the body solders of the automotive industry; they serve to fill, repair and streamline the flow lines of oceangoing vessels. Application conditions for these steel-filled pastes are somewhat rigorous, including the following considerations, after thorough cleaning and degreasing of exposed surfaces:

- Gap filling of altered contours may require materials to be built up to a depth of several inches; worn or eroded parts such as fairwaters may be preserved (Fig. 3-12).
- Physical properties must not significantly change with aging or on contact with sea water; this characteristic requires extensive laboratory and field testing. Shrinkage and age-hardening should be carefully observed.
- Expansion characteristics are concerned with the material's reaction to freezing water and to high ambient temperatures. Although not too severe a thermal cycling, it is possible for a resin formulation to match more closely the expansion characteristics of the steel hull. Inserts of glass or graphite fabrics will reduce the expansion coefficient of epoxy pastes.
- Vibrations due to many causes, from the engines, the movement through water, and the turbulence of high seas, impose strong adhesion requirements. With the added problem of high impact blows from floating objects, a high impact resistant bond is necessary. Hull smoothing cements may fracture and separate from the metal, and in the test procedures outlined by the U.S. Navy Department a calculated attempt is made to pry off $1/16$ in. thick layer hull smoothing cements after they have been fractured by failing ball impact against the materials bonded to steel. When cracks appear, the thick coatings in the immediate vicinity must not be capable of separation by driving in appropriate wedges. Adhesion, particularly for thick coatings such as metal-filled hull smoothing cements, is dependent on the resin binder, its hardener, and modifying agents which offer flexibility. A slight alteration in the formulation may be advantageous since if greater toughness is exhibited, greater peel strength will develop, together with the ability to accommodate vibrational stresses.

Adhesive pastes which are too soft, because of the plasticizer, tend to be lower in chemical resistance. Whatever chemical modification has taken place (usually through polyester, polyamide, polysulfide or polyurethane addition), the increase in toughness and flexibility should be permanent. Otherwise, em-

68 METAL/POLYMER COMPOSITES

Porous Steel Fairwater.

Figure 3-12. Porous fairwater filled with epoxy paste.

brittlement and loss of adhesion, caused by more cross-linking of polymers or traces of volatile compounds, will lead to unsatisfactory service.

Another consideration that arises in the use of hull smoothing cements, is the use of copper-oxide additions to the formulations for the development of anti-fouling properties in contaminated water. Some discussions of anti-fouling compounds in metal/polymer composites is given in Chapter 5 on coatings.

Tank Sealant. A more complete picture evolves when the subject of tank sealant and protection is presented in conjunction with metal-filled pastes. This topic is of major importance to the chemical and food processing industries, but the industries find limited use for metal-filled pastes for external patching and sealing (e.g., over leaky or exposed rivet heads). Care must be exercised in the selection of materials in the paste. The materials must not be adversely affected by the environment to which they are exposed. The preparation of coating and laminating compounds from resins and inert pigments becomes necessary, and such compounds should be applied over metal-filled pastes if they are used for sealing. Adequate rubber-like "O" rings have been essential in the design of some metal tanks containing volatile fuels. This was dramatically demonstrated in the space-shuttle disaster of 1987.

MISCELLANEOUS METAL PASTES

Solders. Most miscellaneous pastes developed from powdered metals have been prepared with the addition of organic, oily compounds such as lubricants and waxes. These efforts reflect the early pastes developed during the beginning of the twentieth century. In more recent years, the availability of resinous binders has prompted new developments. A good example is the fusible powdered metal paste used for soldering purposes. Rosin and rosin esters, acrylic and methacrylic esters, epoxy resins, polyesters, etc., are among the nonaqueous organic liquids which contribute to improved adhesion. Non-slumping, inorganic salt-free, anhydrous, non-corrosive, powdered solder metal paste was developed around 75 to 95 parts of powdered tin-containing solder, which melts below 330°C. The powdered metal particle size was not substantially larger than 100 mesh.[24, 25]

Disk Brakes. Compositions for disk brakes for automobiles continue to undergo developments to attain improved wear resistance and reduced noise levels when braking pressure is applied. The presence of bronze powder (Cu/Sn—90/10), phenolic resin binders, and steel and graphite fibers were found to be acceptable.[26]

Resin-Impregnated Sintered Parts. The development of sound absorbing materials, for auto exhaust mufflers which have good corrosion resistance has been the subject of a number of Mitsubishi patents. Inorganic compounds were preferred as the binders for high temperatures. Porous sintered nickel/chromium alloys were sintered and spray coated with aluminum phosphate and aluminum oxide as the binder[27]; glass frits were used in another application[28]; aluminum phosphate and calcium phosphate were used in another application[29]; and methyl phenyl silicone polymer in a further example.[30]

Metal Powders in Pastes and Paints—Storage. The storage stability of metal powders in combination with organic liquids and polymers should be considered. Some powdered metals such as zinc have been known to form gaseous by-products during storage and negate the usefulness of the product. It is always prudent to perform accelerated tests on all combinations of materials, until sufficient experience has accumulated and remedial measures instituted. The variables involved in the preparation of composites of liquid polymers and metal powders are extensive and laboratory evaluations are essential before such mixtures are distributed to customers. In some instances, the metallic powders are placed in a separate container and mixing is commenced immediately before their application.

INTERACTIONS BETWEEN METAL PARTICLES AND POLYMER PARTICLES

Germane to studies of combinations of metal particles and polymer particles are the interactions that may occur among diverse materials, one of which is basically crystalline in nature, and another which is fundamentally amorphous. The complexity of the relationships grows when it is established that amorphous phases and crystalline phases may exist in solids of both materials.

There is mounting evidence that unreported reactions between polymers and metals do occur, and information on these reactions will allow one to anticipate factors such as instability and degradation. The author has conducted many experiments on combinations of metal powders and polymer powders. These experiments were based upon the premise that when powdered aggregates are thoroughly combined, opportunities are increased for observing untoward effects. In finely divided powders, interactions between metals and polymers are magnified and reveal much more than the application of a polymer coating on a metal surface, or an adhesive application between two substrates.

Structural analysis of interfacial regions of solids is a crucial aspect of materials science because the nature of the interface affects many factors such as solid state electronic parameters, catalysts, oxidation, and corrosion processes. In such analysis, the writer has employed an analytical procedure which has lead to the discovery of unreported interactions between metals and polymers. The subject is raised here because it focuses upon combinations of finely divided metal particles and polymers. These react together at temperatures below the melting point of the metal and before the apparent decomposition of the polymer. Unlike the chemistry of many metallo-organic compounds, the tests illustrated (Figs. 3-13 and 3-14) examine polymeric materials and finely divided metals over a broad range of temperatures, plotting exothermic and endothermic reactions at a specific temperature gradient.

In some instances the apparent decomposition temperature of the polymer is lowered, and in others the thermal performance of the polymer is improved after its interaction with the metal. In most cases, oxidative processes are involved when analyses are conducted in air. Early investigations of interactions of metals and polymers in the presence of argon, as an inert gas, demonstrate similar data. It is suspected that partial thermal decomposition of the polymer has led to the presence of active sites which are capable of reaction with metal surfaces.

The fact that increasing evidence has been accumulating on the influence of specific metallic ions on polymer behavior at high temperature is of significance to metal/polymer composite research. The techniques noted in this chapter reveal some of the anomalous results that are encountered. In real-life applications, there is concern for the stability and strength of polymer systems in the presence of metals or combinations of metals, particularly at higher temperatures.

Typical charts are shown for equal volumes of metal powders and epoxy polymer and powdered metals with polyvinyl chloride. Different mesh sizes of finely divided powders were also studied, as were different ratios of metal/polymer combinations, and different rates of heating. Results were determined in atmospheric oxygen and also in an inert argon atmosphere. As a consequence of this examination, the apparent decomposition of the subject polymers are reported here for equal volume ratios of polymer and metal. I have referred to the apparent decomposition principally because there is evidence of other reactions between the metals and the subject polymers. These reactions appear to have occurred, on the evidence of sharp exotherm spikes, between activated sites on the polymers and the metal surfaces.

72 METAL/POLYMER COMPOSITES

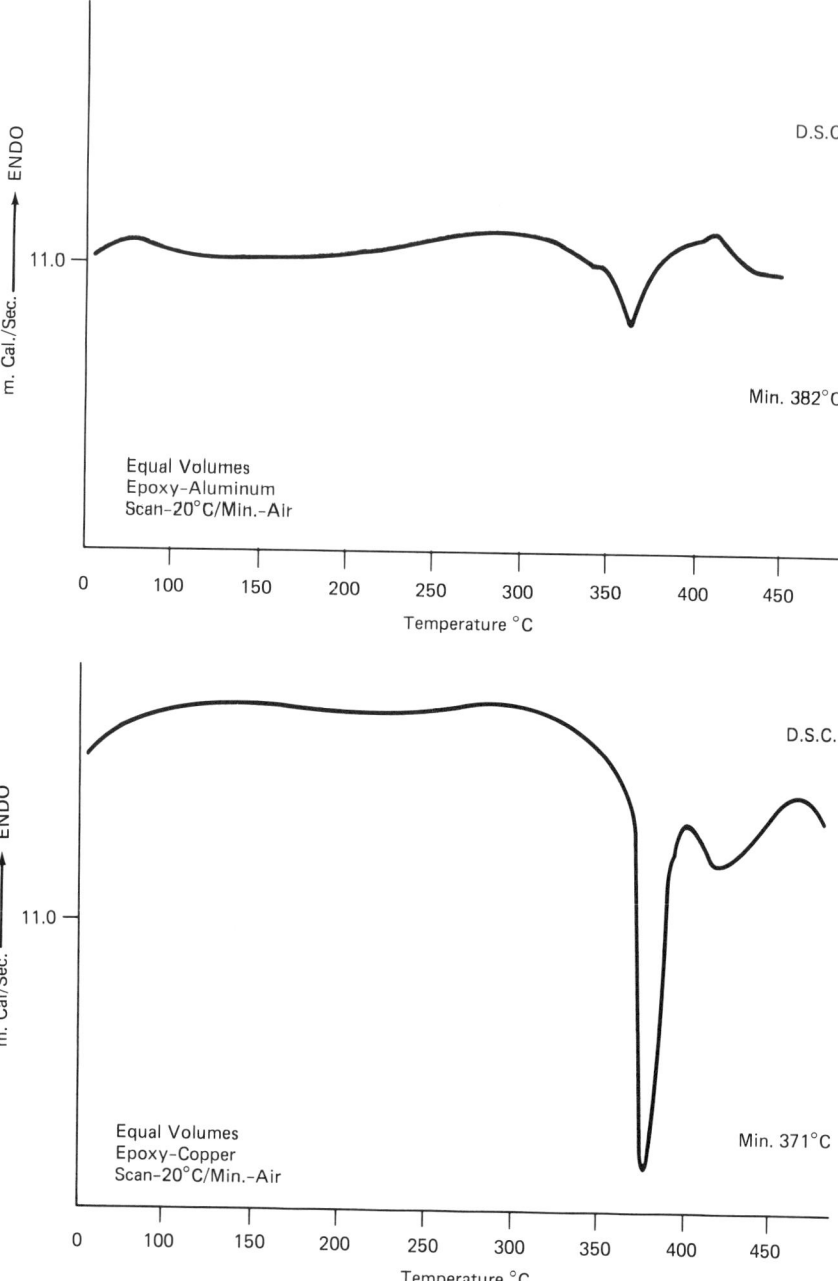

Figure 3-13. Reactions between epoxy resin and metals.

MOLDING AND CASTING OF METAL/POLYMER COMPOSITES 73

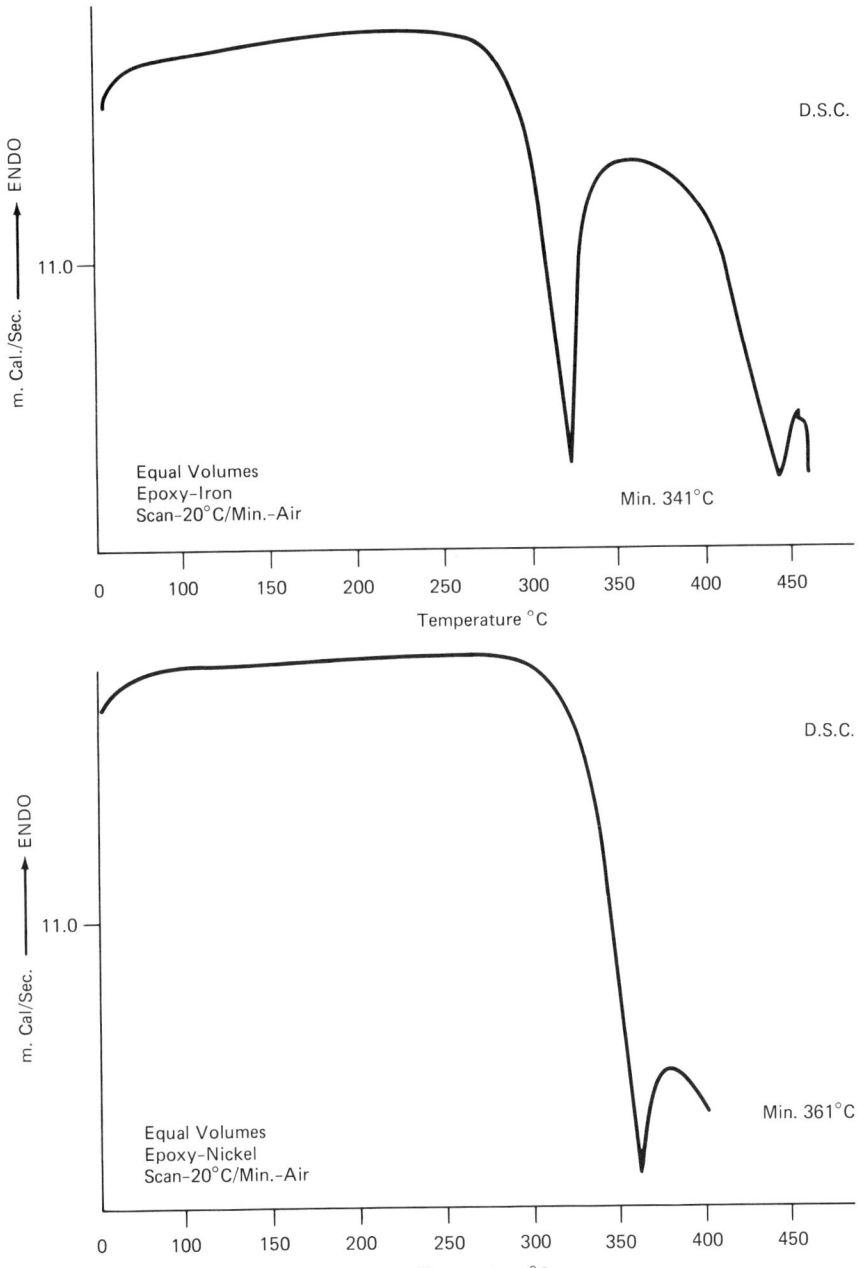

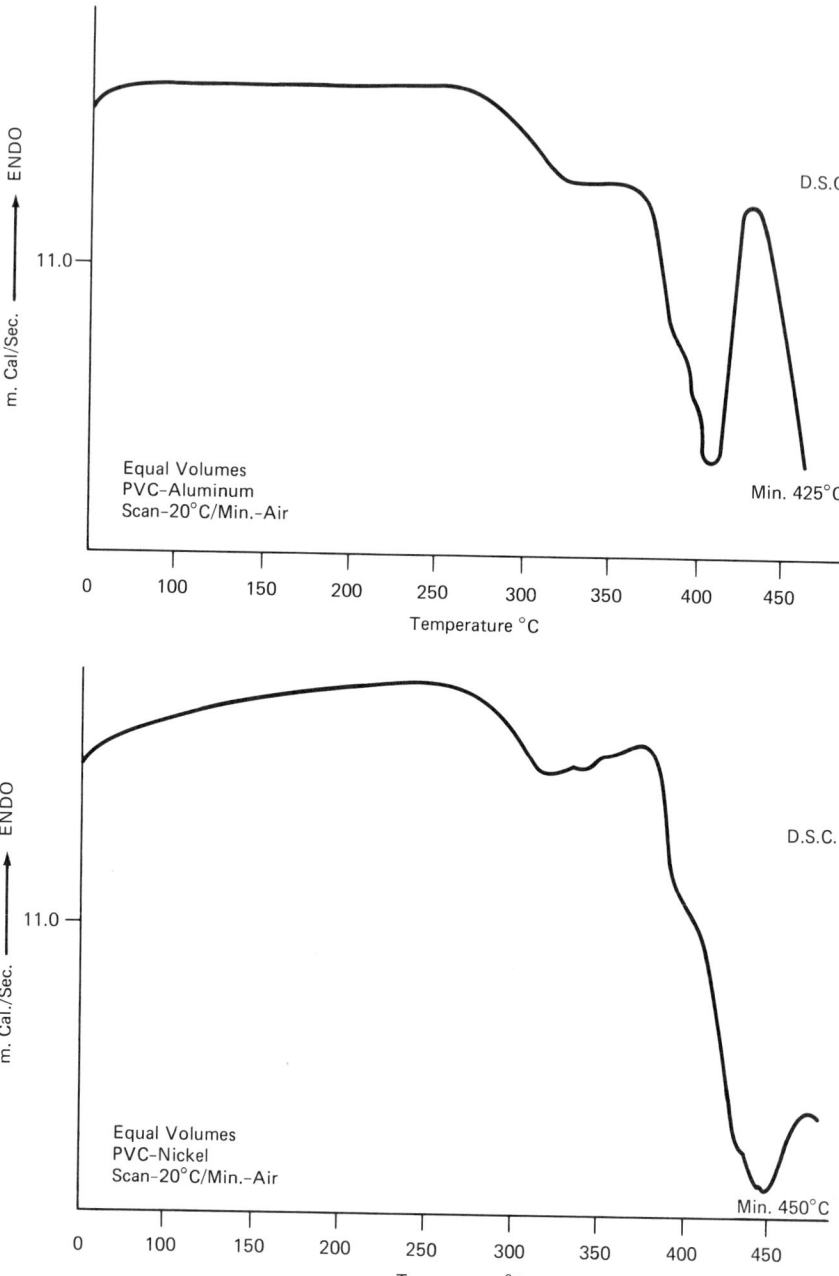

Figure 3-14. Reactions between PVC and metals.

MOLDING AND CASTING OF METAL/POLYMER COMPOSITES 75

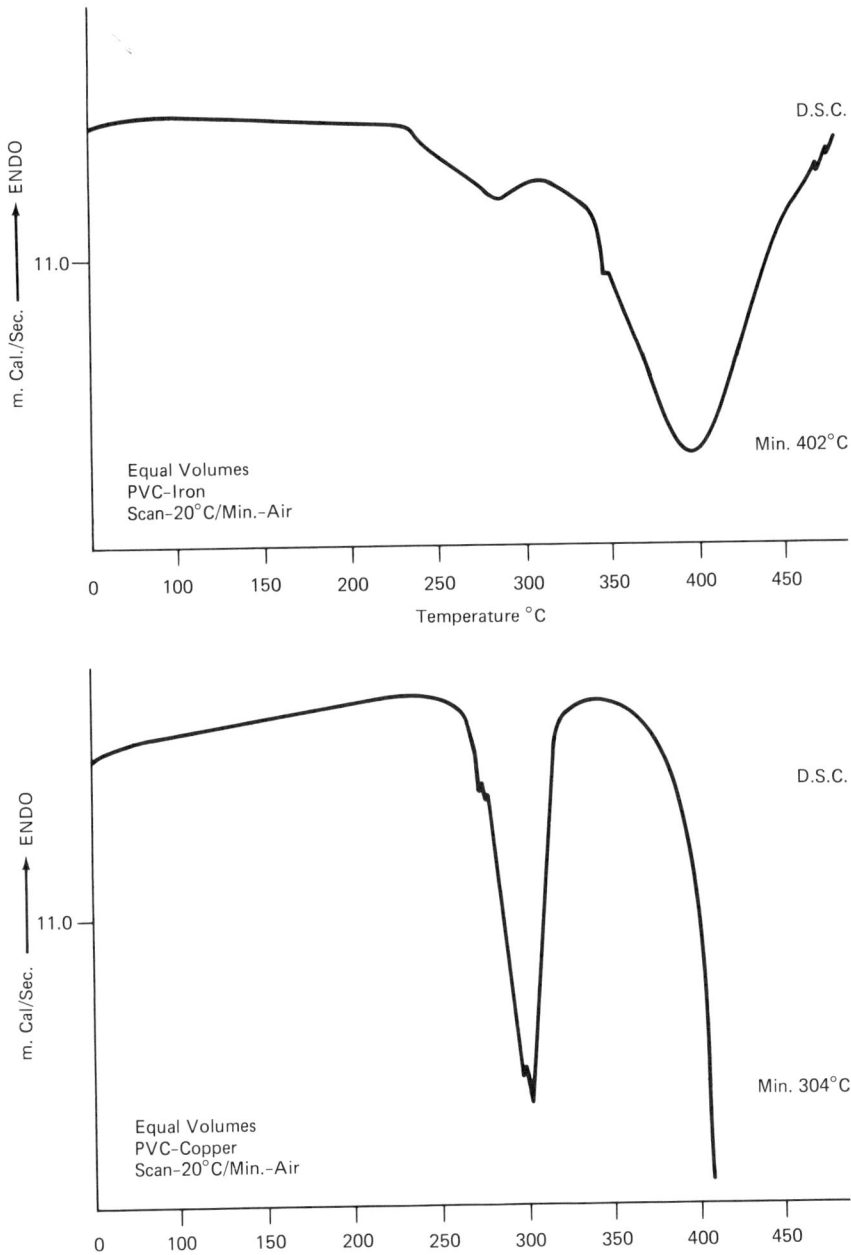

REFERENCES

1. Katz, H. S. and Milewski, J. V., *Book of Reinforcements and Fillers for Plastics*, New York: Van Nostrand Reinhold, (1978–1987).
2. Delmonte, J., *Plastics Molding*, pp. 493, New York: John Wiley, (1952).
2A. *Powder Metal Design Guidebook*, Princeton, NJ: Metal Powder Industries Fed., (1983).
3. Krishnamurthy, V. and Kamel, I., *S.P.E. Annual Conference Antec*, p. 514, (1987).
4. Raj, R., *A.C.S. Polymer Materials*, 57: 537, (Fall, 1987).
5. Anon. *C. and E. News*, 66: 30, (June 6, 1988).
6. Plummer, R., *Powder Metal Int.*, No. 17: 141, (1985).
7. C.A. 107-135, 305b Hirabayashi, JP62, 109.587, (May 21, 1987).
8. C.A. 100-17, 901A Toshiba Chemical Products, JP58, 220.714, (December 23, 1983).
9. Anon. *C. and E. News*, 64: 20, (February 2, 1986).
10. *Creating With Metal Powders*, p. 19, Riverton, NJ: Hoeganaes Corp., (1978).
11. Crook, P. and Rivers, R. D., *Metal Injection Molding*, p. 41. Princeton, NJ: Metal Powder Ind. Fed. (1987).
12. Roll, K., *Processing Powder Metal Conf.* 43: 4, 324, Dallas, Texas, (May 1987).
13. Skah, S. and Hunn, A., *Plastic Engineering*, 43: 33, (November 1987).
14. Carlson, C., U.S. 2,297,691 (October 6, 1942) and *Atlantic Monthly*, 287: 64, (October 1986).
15. Jaray, F., *Wire Reinforced Plastics*, Washington D.C.: SPI Reinforced Plastics Division, Section 7A, (1967).
16. MacNeal, K., *Encapsulants For Integrated Circuits*, S.P.E. Antec Proceedings, p. 358, (1987).
17. Trenkler, G. and Stoeckler, H. to Texas Instruments Co. U.S. 3,952,116, (April 20, 1976).
18. Stinger, H. to DuPont, U.S. 3,900,654, (August 19, 1975).
19. Delmonte J., *Evaluating Plastics Tooling Materials*, S.P.I. Tooling Seminar, Purdue University, (June 12, 1975).
20. Delmonte, J., "Epoxy Resins for Laminated Tooling," *Materials and Methods*, New York: Reinhold Publishing Co., (August 1954).
21. May, C., *Epoxy Resins*, New York: M. Dekker, (1986).
22. Peerman and Floyd, *S.P.E. Symposium*, Minneapolis, (October 21, 1958).
23. Delmonte, J., "Epoxy Pastes Prove Versatile," *Plastics Technology*, (August 1956).
24. Hwang, J. to SCM Corp., U.S. 4,619,715, (October 28, 1986).
25. Hwang, J. to SCM Corp., U.S. 4,557,767, (December 10, 1985).
26. Hitachi, C.A. 100,1373h, J.P. 58,136,676, (August 13, 1983).
27. Mitsubishi Electric, C.A. 100-25196g, J.P. 58,132,289 (August 6, 1983).
28. Mitsubishi Electric, C.A. 100-25197e, J.P. 58,129,492 (August 1983).
29. Mitsubishi Electric, C.A. 100-25198s, J.P. 58,129,491 (August 1983).
30. Mitsubishi Electric, C.A. 100-25199t, (August 1983).

4

ELECTROCONDUCTIVE POLYMER/METAL COMPOSITES

Electrically conductive nonmetallic materials have been in the forefront of scientific research in recent years. The development of high transition temperatures for superconductivity in ceramics has won high acclaim. Equally important are the achievements in obtaining good electrical conductivity in polymers. This chapter covers the principal techniques for attaining good electrical conductivity in polymers:

- Synthesis of electroactive polymers and use of dopants and charge transfer salts
- Deposition of metallic elements in polymers by thermal decomposition of specific metallo-organic compounds
- Metallic ion implantation techniques
- Compounding of polymers with conductive metal powders and carbon black
- Vacuum metallization of polymers is discussed in Chapter 5.

Metal powders with polymers have attained significant commercial success and are discussed at length in the second half of this chapter. Not only are fine metal powders used but also metal coated fibers, metal flakes, and metal coated inorganic particles—contributing a large representation of metal/polymer composites.

The electrical conductivity (and its reciprocal electrical volume resistivity) vary by more than 20 orders of magnitude, as noted in the accompanying chart (Fig. 4-1) adapted from Reference 1. The range of conductivity for some recent organic materials is shown on the right hand side of the diagram. On the following pages is presented an overview of developments in conductive and semiconductive organic polymers.

Electricity flows by the movement of electrons, and within the atom the movement of electrons takes place between discrete energy states called bands. The number of atoms in a solid can be approximated by 10^{23} atoms/cm^3, and the net effect of assembling a large number of interacting atoms is to establish

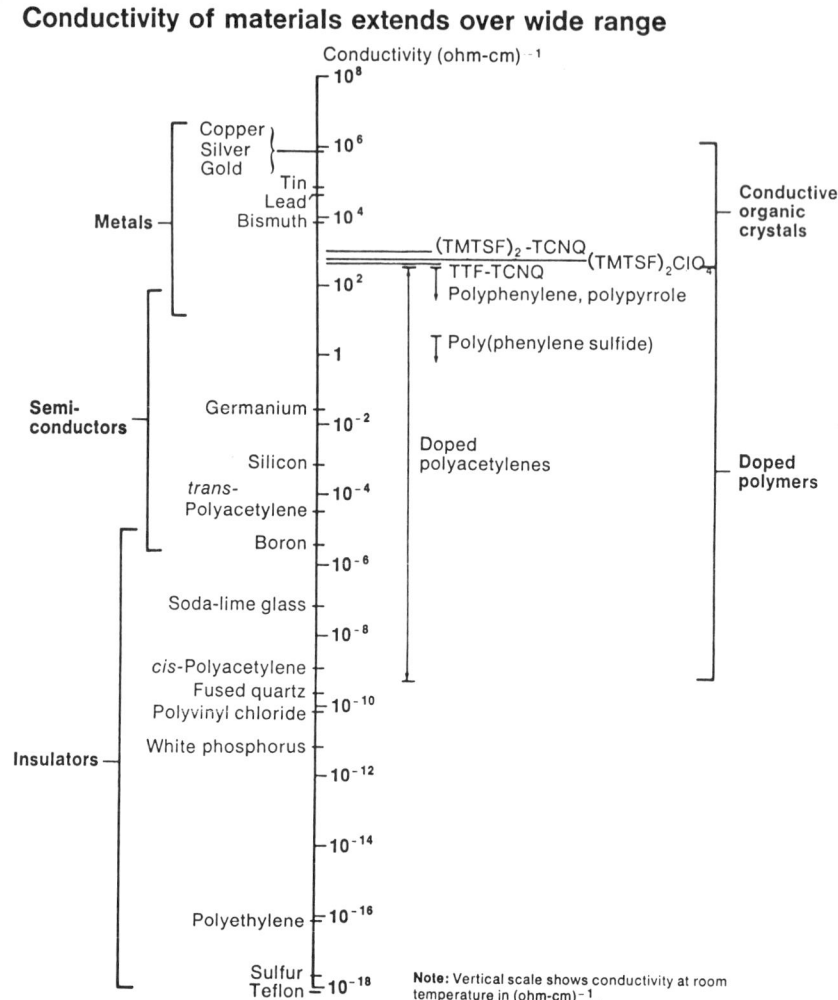

The electrical conductivity of materials varies by more than 20 orders of magnitude between a good insulator and a metal. The range of conductivity at room temperature found for some organic materials is depicted on the right-hand side of the diagram. At a temperature of about 1 K, $(TMTSF)_2ClO_4$ becomes a superconductor (zero resistance). The classifications of insulators, semiconductors, and metals on the left-hand side of the drawing are only approximate. For example, a material with a room-temperature conductivity of 100 $(ohm-cm)^{-1}$ could be a metal, but it could also be a semiconductor. A decision can be made only after its electrical transport properties have been studied as a function of temperature and structural studies performed

Figure 4-1. Chart of volume resistivity. Courtesy of Dr. Cowan and Dr. Wigul and A.C.S. *Chemical & Engineering News.*[10]

bands of practically continuous energy levels, separated by gaps where no electrons exist. The bands in solids may be filled, partially filled, or empty. If the band containing valence electrons is filled, it is referred to as the valence band, and the next higher band is the conduction band. A good electrical conductor has a conduction band partially filled or else there is some overlapping of the valence electrons that may acquire energy from the electrical field to participate in the conduction.

Insulating materials have filled valence bands and a large energy gap separating the conduction band from the valence band. Semiconductors have filled valence bands, like an insulator, but a small gap to the conduction band.

An important influence on the electrical conductivity of a material in an electric field is the scattering of the electrons by lattice vibrations. The conductivity of a metal tends to increase as temperature decreases (many metals reach superconductivity as absolute zero $-0°K$, is approached), and decrease with a rise in temperature. On the other hand, with electroactive polymers more electrons are promoted into the conduction band at higher temperatures and conductivity increases. There is a metal semiconductor transition—usually below room temperature, and at further temperature increases, conductivity decreases, behaving more like an elemental metal.

TRENDS IN ELECTROACTIVE POLYMERS DEVELOPMENT

In 1975 MacDiarmid brought Shirakawa from Japan because of his development of a silvery film of polyacetylene. In 1976, Shirakawa, working with Heeger and MacDiarmid discovered that iodine added to the polymer increased conductivity more than 10 orders of magnitude (over a billion-fold). Iodine is a p-dopant, the molecule gets a net positive charge. Prior to this research, there had been a short lived interest in a polymer of sulfur and nitrogen $(SN)_x$. Though first synthesized in England in 1910, it was "rediscovered" in the 1950s, and in the mid-1970s Labes of Temple University showed that the polymer $(SN)_x$ conducted electricity at room temperature. Superceded by other polymeric conductors, it has not received much attention.[2]

It should be noted that the number of organic electrical conductors has been growing and that conductivity is influenced by organic charge transfer compounds.[3] The first stable electrically conductive organic material was synthesized at the duPont Company in 1960. The compound was 7, 7′, 8, 8′—tetracyano-p-quinomethane, usually abbreviated TCNQ. Lupinski, of General Electric Company, attained conductive polymers impregnated with TCNQ.[4] Rembaum, of Cal Tech, also commented at the same time that the mechanism of conduction through the complex TCNQ is electronic rather than ionic, involving the moving of electrons along conductive bands.[5] It was not until 1972 that Cowan and Ferraris, at John Hopkins, found that single crystals of the salt TTF-TCNQ showed metal-like conductivity.[10] Organic radical cation salts derived from electron donor tetrathiafulvalene (TTF) have opened new areas of chemistry and physics.

Vibrational spectroscopy has proven useful in the study of organic charge-transfer compounds. The studies have provided information on the nature of the interaction between the conduction electrons and molecular vibrations. Tetrathiafulvalene, and related molecules, and the salts formed with I_3^-, $AuCl_2^-$, AuI_2^-, IBr_2^-, ClO_4^-, were among those studied. On the reflectance spectra for the charge transfer salts, four different structures are involved.[6] Packing diagrams for several of the molecules in the environment of tri-halide anions are illustrated in Reference 6. These data are reported here as representative of the basic studies underway to elucidate the practice of electrical conductivity in organic compounds.

Heeger has reported on techniques for studying charge transfer doping reactions in conducting polymers. These techniques have made it possible to monitor the kinetics of the charge transfer reaction. The dominant electronic excitations in conducting polymers are coupled to nonlinear distortions of the organic molecule.[7]

In analyzing design trends for polymers as electronic materials, Duke cites the established areas of insulation for communications and electrical machinery, the electronics industry (printed circuits and wiring boards, photo-resists), the printing industry (photo-lithographic plates), and the electrophotographic copier industry.[8] He further outlines salient difference between the physics of electronic processes in polymers as opposed to those in network semiconductors. The interpretation of fundamental phenomena remains inconclusive, as does the mechanism of the insulator to metal transition, and the mechanisms of charge conduction. Continued study is stressed for polyacetylene with new forms of iodine—doped polyacetylene $(CH)_x$ which possesses conductivities of 30×10^3 S/cm.[9]

Polyacetylene has been a favorite candidate for research on the electrical conductivity of organic polymers. There are two principal isomers (cis and trans) which are depicted in Figure 4-2. The influence of isomer content on the electrical conductivity of iodine-doped polyacetylene was reported by Kusy.[11] The cis-isomer was the more conductive material:

CIS ISOMER (PERCENT)	CONDUCTIVITY (S/CM)
98%	500–700
74%	105–250
24%	50–80
7%	3×10^{-4}

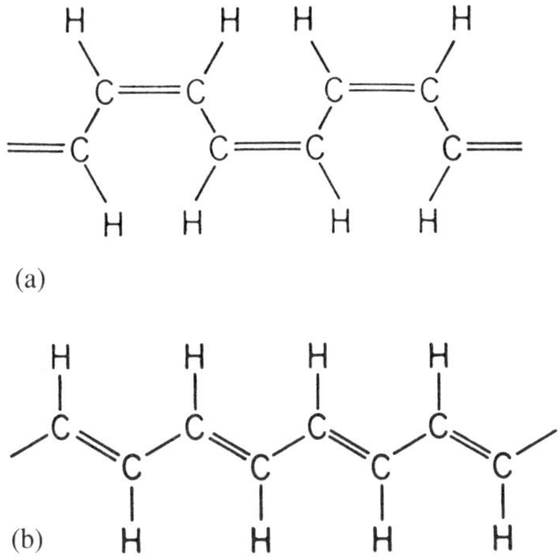

Figure 4-2. The chemical structure of polyacetylenes: a. cis isomer, b. *trans* isomer.

Studies on electrically conducting polymers were reviewed in reports to a division of polymer chemistry in an ACS-Symposium in 1982. A summary of experimental results at that time was reported as follows[10]:

Table 4-1. Influence of Some Dopants on Conductivity of Polymers

POLYMER FILM	DOPANT	CONDUCTIVITY S/CM*
Polyacetylene	Arsenic pentafluoride	1200
Poly-p-phenylene	Arsenic pentafluoride	500
Polypyrrole	Iodine	100
Polyphthalo - cyaninesilorane	Iodine	1.4
Copper		6×10^5

*Conductivity is the reciprocal of resistivity = $mho^{-1} cm^{-1}$
Conductivity is also expressed as S/cm, where S is the symbol for Siemens.

SYNTHESIS OF ELECTROACTIVE POLYMER DEVELOPMENT

Activity in the development of electroactive polymers has proliferated greatly in recent years, and many scientific papers have appeared on this subject. The book edited by Skotheim has presented an up-to-date account of the latest (1986) developments in electroconductive polymers and is suggested as a comprehensive reference.[12] Examples of metal/polymer composites are not examined in any detail in that publication, but considerable attention is given to the subject of electronic transport in polymers. In the following pages a few examples are presented to demonstrate recent activity in the synthesis of electroactive polymers. Some of the metallo-organic polymers cited would be in the category of composites because of the formation of elemental metals within their structures. The latter part of this chapter concentrates primarily on metal-filled polymers. Other publications of note are References 13 and 14.

Several recent papers bearing on the synthesis of electroactive polymers are noted below.

1. Baughman divided highly conducting polymers (conductivities greater than 1 S/cm for unoriented polymers) into two classes.[15] Class I comprises those polymers which do not undergo irreversible dopant-induced chemical changes to become highly conductive. Polyacetylene and poly (p-phenylene) are cited as examples. Dopant-induced chemical changes, in addition to charge transfer, might occur to a limited extent for class I polymers. On the other hand, class II polymers become highly conductive as a consequence of the polymer backbone resulting from dopant-induced reaction. As an example, poly (m-phenylene) with A_5F_5 dopants, forms interchain phenyl-phenyl cross links at the metal ring positions.[15]

2. More recently, Jenekhe has described a group of electroactive polymers where the backbone is selected from a combination of aromatic and quinonoid bonding structures. These have been selected from a class of five numbered heterocyclic rings consisting of furans, pyrroles, and thiophenes.[16] Smaller band gaps are claimed for the electroactive polymer incorporating both aromatic and quinonoid groups, than are obtained with solely aromatic geometry in the polymer structure.

3. Polymers of N-substituted carbazoles and substituted benzaldehydes were prepared in the presence of sulfuric acid. They were doped with charge transfer acceptors such as Br_2. Resistivities of 0.3 ohm/cm were reported.[17]

4. High temperature treatment of polyoxidiazole (POD) yielded a flexible film of high electrical conductivity. Lattice spacings of oriented graphite crystallites were determined at 3.54 angstroms. Resistivities of the order of 10^{-5} ohms per micrometer were reported.[18]

5. Electrically conductive polymer blends are being applied as a surface layered to insulative non-porous polymers to increase the electrical conductivity of the polymer base. Particularly attractive for incorporation into conductive polymer blends are conductive polymers of polypyrrole, polyaniline and their analogs, which will exhibit good environmental stability. An electrically conductive surface layer, from 1 micrometer to 1 mm deep may be attained from treatment with solution blends of pyrrole and polyvinyl chloride, applied to a host film or fiber of polyamide, polyvinyl chloride, polyurethane, or polycarbonate. Upon evaporation of the solvents a catalytic amount of acid is added with an oxidant, iron trichloride in hydrochloric acid. The pyrrole polymerizes to polypyrrole and turns black. Similarly when aniline monomer and its oxidant ammonium persulfate are applied to the host polymer, a dark colored conductive surface is obtained.[20]

6. Poly-o-methoxy aniline (poly-anisidine) has provided a soluble conducting polymer. This compound was chosen for examination because of the effect the methoxy group may have on chain solubility. The orthoisomer was selected because a linear chain polymer was more likely. The o-anisidine was polymerized electrochemically under acid conditions (1 molar HCl). The polymer has an electric conductivity of 3 S/cm. The film is stable in air, and exhibits no change in conductivity after several months.[21]

7. Electroactive polymers were prepared by electrochemical doping of various quinoline polymers. These polymers consisted of recurrent units of fused nitrogen containing unsaturated heterocyclic ring systems. The dopant consisted of sodium anthracenide in tetrahydrofuran.[22]

8. Conducting polymer complexes consist of ionized polymer backbones with appropriate counter-ions interdispersed between the charged chains. Elsenbaumer lists some of the more frequently used counter-ions.[74]

For acceptor doped polymers (p-type)
BF_4^-, PF_6^-, AsF_6^-, ClO_4^-, etc.

For donor doped polymers (n-type)
Li^+, Na^+, K^+, etc.

The best long term air stability at room temperature has been reported for certain p-type polymers such as polypyrrole, polyaniline, and poly (3-alkyl thiophenes). Tests were conducted upon 7 mm diameter disks (15 to 50 micrometers thick) which were cut from doped pre-cast polymers.

9. An electrically conductive polymer blend is described in which pyrrole and its derivatives are treated with a solution of a chemical oxidant selected from trivalent iron compounds ($FeCl_3$), tetravalent cerium compounds, hexavalent molybdenum or chromium compounds.[59] The host compound was polyvinyl chloride, polyurethane, polycarbonate, or polyamide (nylon staple fiber 30 μm diameter). Conductivity of untreated nylon fiber was low and when treated with the oxidized pyrrole (10% in acetonitrile) the conductivity was as shown in Table 4–2.

10. A recent discovery by Thakur of AT&T has challenged the theory that in order for polymers to conduct, the atoms in the chain must be separated by

Table 4–2. Effect of Treatment on Nylon Fibers[59]

	TREATED FIBERS	UNTREATED FIBERS
Conductivity S/cm	1.3×10^{-3}	$< 10^{-10}$
Stress at failure (psi)	2.5×10^5	1.1×10^5
Strain at failure (%)	54.0	75.0

alternate double and single bonds (as in polyacetylene). He reported that the doping of natural rubber (isoprene) with iodine increased the electrical conductivity of the natural rubber by 10 orders of magnitude.[24] This observation is subject to the doping procedure and sample preparation technique, and may reflect primarily a thin film surface conductivity on a temporary basis.

POLYACETYLENE AND POLYTHIOPHENE DEVELOPMENTS

Several recent developments concerned with polyacetylene and polythiophene are noted below.

1. Attention is called to the effects of chemical doping on the room temperature conductivity of polyacetylene powder pressed into a pellet. The highest room temperature conductivity achieved is 1×10^{-3} S/cm with electron acceptor doping increasing the conductivity by a factor as high as 10^3. More recently, polyacetylene has been synthesized in the form of copper colored polycrystalline flexible films of cis-polyacetylene (at temperatures under $-78°C$), and silvery polycrystalline films of trans-polyacetylene in the presence of $Ti(OC_4H_9)_4Al(C_2H_5)_3$ catalyst. Room temperature conductivity of trans-polyacetylene is reported at 4.4×10^{-4} S/cm while the cis-polyacetylene is typically 1×10^3 S/cm. Improved doping methods permitted the production of a whole family of p-type electrically conductive doped polyacetylene films. Electron acceptor dopants, iodine, bromine, arsenic pentafluoride, and iodine chloride (0.1 to 0.3 mol of dopant per C-H) were applied in a vapor phase to the films.[28]

2. Polyacetylene (predominantly in the cis configuration) is doped with potassium or rubidium salts electrochemically. According to Delannoy, heat treatment at 100°C to 250°C will achieve enhanced room temperature electroconductivity of 300 to 400 S/cm. By shaping the materials prior to doping or prior to heat treatment, considerable versatility is introduced into the process.[23]

3. A new process for the preparation of metal-like, stable polyacetylene has been introduced. The acetylene polymerization was catalyzed by $Ti(OC_4H_9)_4Al(C_2H_5)_3$ in a *silicone oil* reaction medium. Stable, defect-free polyacetylene films, capable of being stretched up to 400% were prepared. When the film had been stretched and doped with iodine, conductivities of 16,000 S/cm were obtained. The polyacetylene films were doped in a saturated solution of iodine in carbon tetrachloride.[25]

4. The electroactive polymers produced from polyacetylene and polypyrrole have been brittle and difficult to mold or to fabricate in practical applications. BASF has claimed changes in these two polymers by adding ring like structures (not identified) into the polymer backbones. These electrically conductive polymers are easy to process and possess adequate flexibility.

5. The dependence of electronic conductivity upon conjugation length for conducting polymer was calculated for heavily doped polymers with equivalent chains. Good agreement was found for I-doped polyacetylene and K-doped poly (p-phenylene).[26]

6. Polythiophene thin films when doped with iodine demonstrated semiconductive properties. Resistivity of $(C_4H_2SI_y)_x$, and iodine doped polythiophene was 10^2 ohms/cm after doping and 10^8 ohms/cm before.[27]

7. The dominating tendency of thiophene monomer is to polymerize into polymer chains with extensive electronic delocalization. This assures high conductivity upon doping (I_2 48 hours at 60°C or AsF_5 warm at 60°C). Pyrolysis shows polythiophene films are stable at 400°C in an inert atmosphere with no pyrolysis fragments below 400°C.[30]

8. A review of polythiophene (PT) and poly (1, 5 thienylene-sulfide) (PTS) polymers showed no terminal groups in the final polymer other than hydrogen. It was surmised that terminal groups interfere with electrical conductivity.[31]

9. A recent paper was presented on the preparation of polyacetylene by the condensed-phase metathesis polymerization of 1, 3, 5, 7-cyclooctatetraene (COT). The dissolution of catalyst into COT provided a means of transforming the liquid into a high quality film.[19] Tungsten carbenes were effective in the ring opening metathesis polymerization. Using ruthenium catalysts, it has been possible to engage in ring opening metathesis polymerization (ROMP) of cyclic and olefinic hydrocarbons such as norbornenes. Novak and Grubbs discovered that the addition of water to the ruthenium tri-chloride catalyst resulted in rapid polymerization of 7-oxanorbornene monomer, yielding higher molecular weight than possible in organic solvent systems. All of the unsaturation in the monomer has been maintained, and the development of good high M.W. films of polyacetylene has been made possible.

DEPOSITION OF METALLIC ELEMENTS BY THERMAL DECOMPOSITION

There has been a continuing interest in the thermal decomposition of metalloorganic compounds to deposit an elemental metal on an insulating surface. In the examples shown below, the objectives were to deposit a continuous metal film for the purposes of electrical conductivity.

1. An early example noted the preparation of a thermosetting phenolic resin solution containing copper formate. When the phenolic resin film was cured at the decomposition temperature of the metal formate, the copper particles contributed to the electrical conductivity, permitting further electroless plating with copper.[29]

2. The heating of a Cu_2O filled polymer substrate to provide conductive paths for producing printed circuits has been proposed. The copper oxide which

was mentioned has a high thermal decomposition temperature. Hydrated copper perchlorate would decompose at lower temperatures.[32]

3. Polymer compositions are being developed by Takakura, in which an organic metal complex is solubilized in a solvent and film is cast upon an inert substrate. Moderate heat is applied to remove the solvent, followed by higher temperature heating which releases metal particles from the organic metal complex. This development of the flexible polymer film with a metallized surface takes place against a casting board, such as a copper sheet. The metal surfaced film stripped from the "casting board" contains a dense aggregation of fine metal particles (1 to 25 μm thick). Thermal decomposition takes place at 175°C, for the example most frequently cited, di-u-chloro-bis (n-methylallyl) dipalladium which had been dispersed in polysulfones. The ratio of polysulfone to palladium was 90-10 to 97-3. Marked lowering of surface resistance occurred for the film to 3.5 ohms/square.[33]

4. The incorporation of bis (trifluoroacetylacetonate) (TFA) copper to polyamic acid prior to imidization produces bi- and tri-layered copper/polyimide films upon curing. Tri-layered films are characterized by a zero-valent copper sandwiched between polyimide films. Increased dopant concentration (copper-TFA) yielded a bi-layered structure, with a copper layer (50 angstroms) being established on the thermal decomposition of Cu(TFA).[34]

METALLIC ION IMPLANTATION

Recent work on ion implantation of polyphenylene sulphide (PPS) was performed on 25 μm films cast on aluminum plates (which served as a heat sink). The conductivity with lithium ion implantation did not change with doses up to 10^{15} ions/cm^2. Higher conductivity was reported at 50 MeV with fluorine and iodine ions, and it was concluded that ion implantation in PPS was dependent on the energy dissipation rate rather than upon total ions.[35] PPS has been a candidate for an electroconductive polymer. The basic polymer is commercially available and thermally stable. After chemical doping with AsF_5, the polymer is amorphous and develops moisture sensitivity.

Controlled doped regions of conducting polymers were achieved by ion implantation. There was a Gaussian-like distribution centered at 1000 angstroms below the surface. By reducing the accelerating energy from 150 to 5 kev, maximum concentration of implanted ions can be brought closer to the surface.[36]

Semiconductors and printed circuit boards have developed specific needs for electroconductive areas. Some of these requirements have been fulfilled by ion implantation techniques which are described more fully in Chapter 9.

PHOTODIELECTRIC ANALYSIS

Examination of the progress and the effects of electron flow through polymers was introduced by Delmonte in some original research in 1947.[61] The technique required the use of transparent polymer models, which were fabricated to simulate cast and laminated materials. At this early date, the materials available to the author for this study were transparent (slightly yellowed) cast phenol-formaldehyde resins and cast polymethyl methacrylate. One of the objectives was to ascertain the effects of the physical shape of conductive electrodes (metals), which were inserted into the polymer structure, upon electron flow under increasingly high voltage gradients. At that time, the subject of metal/polymer composites was not under consideration, though composites were the elements of the study.

Figure 4-3 illustrates the apparatus used in testing the effects of voltage stress upon transparent plastics. A source of polarized light and a polaroid viewing screen permit the observer to witness changes within the material, as the voltage stress is being applied. For observing dark and light fringes, rather than the spread of the colors of the spectrum, a monochromatic source of light is preferred. Specimen preparation required care to ensure that voltages did not produce flash-over on the surface. Distances of the order of 1 cm were used for the spacing of the electrodes. Annealing of test specimens after machining was practiced to eliminate residual strains in the specimens before voltage stresses were applied.

Flat ended metal electrodes and electrodes with slightly tapered points were used as models. Reproduced in Figure 4-4 are the photographic results obtained with cone shaped electrodes inserted into the transparent plastic. Ta-

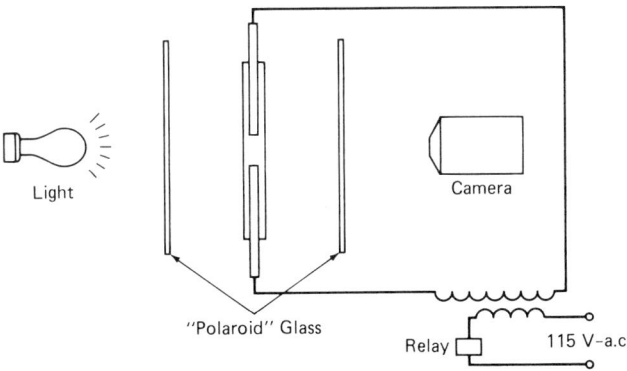

Figure 4-3. A schematic illustration of the apparatus used to test the voltage stress of transparent plastics.

88 METAL/POLYMER COMPOSITES

(A)

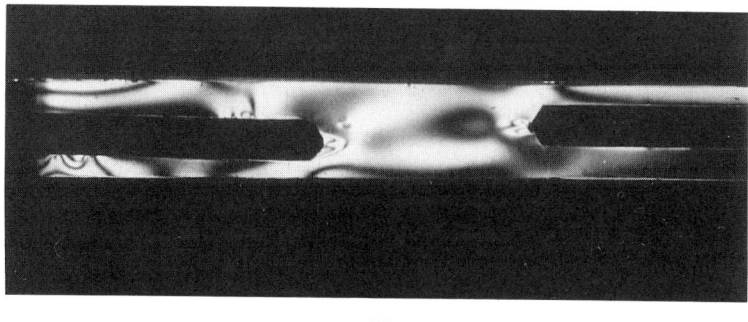

(B)

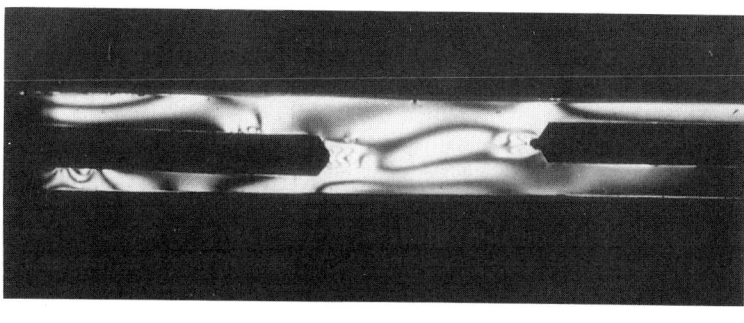

(C)

Figure 4-4. Photodielectric study of transparent (phenolic) polymer under increased voltage stress (at 25°C). A. Before voltage applied. B. At approximately 70% of breakdown voltage. C. After application of breakdown voltage. Tapered brass (3 millimeter diameter) electrodes.

pered electrodes showed the onset of dielectric anomalies at voltage gradients as low as 30% of ultimate dielectric strength. The changing patterns of stress, as viewed in polarized light, are presumed to be a consequence of the gradients in electron flow that take place in the polymer. It is possible that this technique may be applicable to study the effects of low concentrations of dopants in transparent polymers.

ELECTRICAL CONDUCTIVITY IN METAL FILLED COMPOSITES

The study of electrical conductivity in metal filled composites is approached with entirely different interpretations than those required for the study of electroconductivity in polymers. In fact, most references to metal/polymer composites take cognizance of the roles fulfilled by powdered metals, metal flakes, metal fibers, and metal coated organic and inorganic particles, discussed earlier in this chapter. Electrical conductivity in metal filled composites becomes a function of volumetric relationships between the conductive metal particles and the insulating matrix or insulative particles. Under these conditions, the size and the shape of the metal particles and their surface chemistry become significant, as does the rheology of the mixing operations. In general, the metal particle sizes are 0.1 μm (100 nm) and larger. Hence consideration of electrical conductivity in metal filled composites focuses upon the degree of physical contact between the conductive metal particles. For electro-conductive polymers, the analyses study electron flow between overlapping conduction bands of adjacent planar molecules and the stacking sequences of specific polymer derivatives which possess charge transfer salts.

The volume effects of conductive particles dispersed in an insulative polymer matrix are related to a phenomenon called the percolation threshold.[37] Determinations of electrical conductivity indicate a steep rise of conductivity within a narrow range of concentration, establishing a percolation threshold. In the tests performed upon water in oil, a threshold is reached at which an infinite path exists through the aggregate droplets, and conductivity increases sharply.

Examination of conductive metal particles in a polymer matrix shows the marked dependence of electrical conductivity upon the volume percent of the metal. When the volume of metal particles is small, there is very little change in the conductivity due to the large gaps separating the conductive metal particles. An exception occurs when the additions are made to a liquid polymer and the higher specific gravity metals settle to the bottom of the container. When there is thermal blending, and uniform distribution is achieved and maintained by cooled, solid thermoplastic resin, the loss factor and dielectric constant measurements will detect the dispersed metal particles.

Bigg concentrates upon the significance of the close proximity of metal particles and actual physical contact,[39] and develops relationships between resistivity and volume concentration. He comments that the ability of an electron to jump a gap under a given voltage field increases exponentially with decreasing gap size. Gaps as large as 10 nm can be jumped and the process is referred to as hopping. An activation energy is necessary to promote the flow of electrons from one band to another. The presence of fibrous conductive fillers will reduce the isolation of the metal particles and facilitate good electrical conductivity in the metal/polymer composite. The examples cited below, show that this condition has been observed for many formulations, also illustrated is the electrical conductivity (S/cm) versus the volume fraction of aluminum flakes in various polymers. Tests conducted at Battelle showed aluminum flakes, originally with an aspect ratio of 50:1, were fragmented during injection molding of the polymers. Three polymers are shown: polypropylene (■), polycarbonate (△), and ABS (○) (Figure 4-5). The large increase in conductivity occurred abruptly as physical contacts were established between flake fragments.[39]

Other information of basic interest is the relationship between the fiber aspect ratio and the minimum conductive filler concentration required to produce a composition having a conductivity greater than 10^{-2} S/cm. The concentrations of conductive fillers to produce a continuous network decreases very sharply with increasing aspect ratio. In practice, mixtures of conductive fibers with conductive fillers such as stainless steel, or nickel coated graphite fibers, have established significant cost savings in the use of costly conductive filler particles.

The sharp upturn in volume conductivity when volume fraction reaches a critical value, as reported above for small metal particles, may also apply to fine, chopped fibers. When polypropylene was loaded with 0.5 to 1.5% volume of stainless steel fibers, the resistivity rose rapidly.[38]

Powdered silver has an important commercial market for the preparation of electrically conductive adhesives for electronic devices using an epoxy resin base. The volume resistivity of the cured adhesive as a function of the volume concentration of the silver powder shows an abrupt change at about 35% volume loading, from an insulator to a semiconductive to a conductive medium. This reflects the uniformity of particle size (325 mesh) as well as the concentration of physical contacts at the silver surfaces. Silver filled conductive epoxy adhesives have been effective in the bonding of silicon chips to electronic circuits, though tendencies of migration of silver into the gold coatings are reported. Gold has been the most effective filler and the most costly.

Milewski investigated the effect of the fiber aspect ratio on the maximum packing fraction of random distribution.[40] The concentration of fibers required to produce a continuous network decreases sharply with increasing aspect ratio (length to diameter) of the conductive fiber. Bigg illustrated the relationship

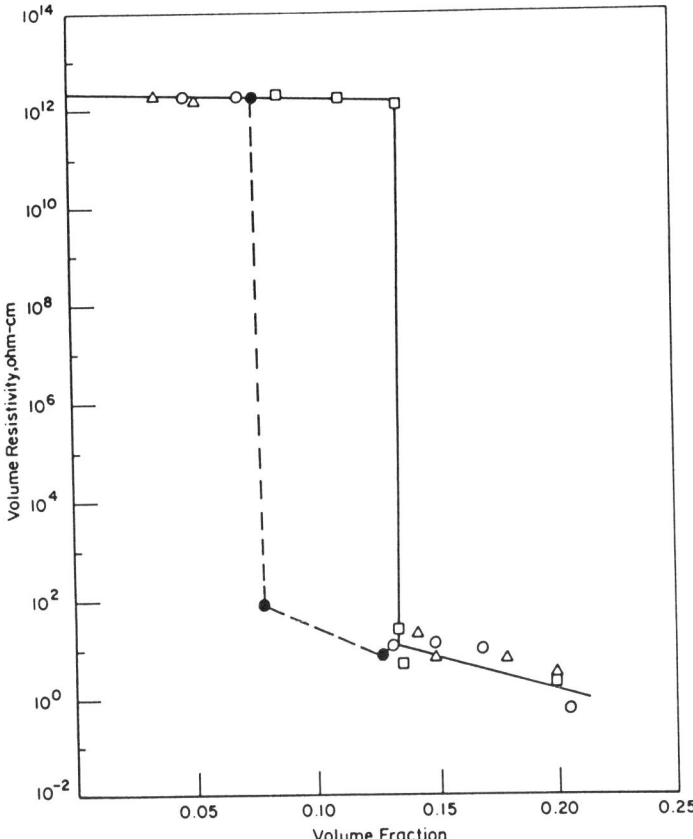

Figure 4-5. Volume resistivity vs. volume fraction of aluminum flakes in various polymers. Open symbols represent injection-molded composites with 50:1 aspect ratio flakes. Filled symbols represent compression-molded composites 16:7:1 aspect ratio flakes. (□, ●) polypropylene, (△) polycarbonate, (○) ABS. Courtesy of D. Bigg,[39] and M. Dekker Publisher.

between the fiber aspect ratio and minimum conductive filler concentration to produce a composite having a volume resistivity below 100 ohm-cm. This data is reproduced in Figure 4-6 from Reference 39.

Because the subject of thin metal films on plastic substrates arises so frequently when examining microelectronic devices, it is desirable to have information on the surface resistance of a metallic film for thicknesses of 20 to 100 angstroms. Figure 4-7 reproduces data by Liao[73] for thin gold film. For aluminum, copper, and silver films, the changes that take place when the materials

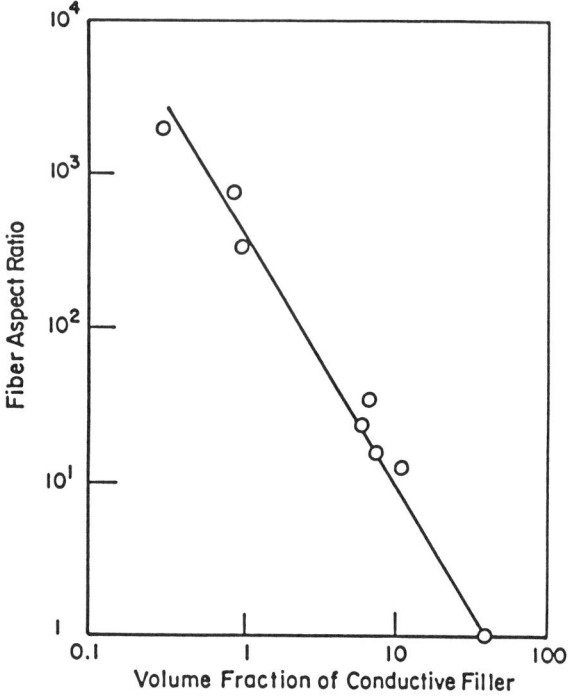

Figure 4-6. Relationship between fiber aspect ratio and minimum conductive filler concentration to produce a composite having a volume resistivity below 100 ohm/cm. Courtesy of D. Bigg.[39]

are exposed to oxygen and other gases must be examined. In all instances, the sealing of the metallic film by a thin polymer coating immediately after deposition, will preserve its integrity.

Table 4-3, taken from standard handbooks, presents comparative elemental properties of conductive metals, which are covered in the reviews of recent publications and patents on metal/polymer composites. Comparative data of (99%+) pure metals, at temperatures of about 20°C are given.

The table shows data on volume resistivity. Surface resistivity (ohms per square) will be affected by environmental conditions and the products of reaction of O_2, SO_2, NO_2, CO_2, H_2O, and miscellaneous gases and fumes upon the metals. When chemical sensitivity is a known factor, protective organic coatings may be applied during the processing of the device. These measures appear among the examples of electrically conductive metal/polymer composites which follow.

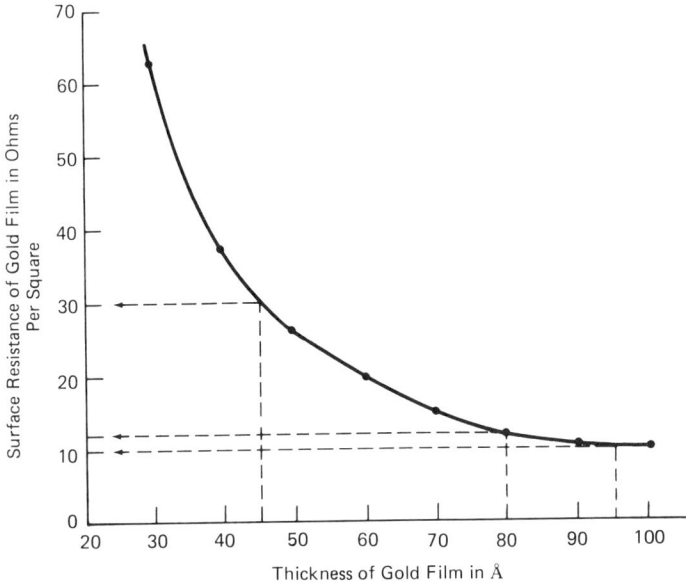

Figure 4-7. The surface resistance of gold film vs. thickness in Angstroms, on plastic substrate, as calculated by S. Liao.[73]

Table 4-3. Characteristics of Pure Elements

MATERIAL	DENSITY (G/CC)	THERMAL CONDUCTIVITY (G CAL/SEC - CM^2/°C/CM)	RESISTIVITY 10^{-6} OHM/CM
Aluminum	2.69	.50	2.16
Copper	8.92	.92	1.60
Iron	7.86	.108	10.0
Gold	19.30	.744	2.4
Lead	11.3	.083	21.0
Nickel	8.9	.140	6.8
Silver	10.5	1.09	1.60

Recent examples of electroconductive metal polymers may be grouped as follows:

1. Polymers plus metal powders which are used as conductive coatings, conductive pastes, or extruded sheets. Electrolytic copper powder is treated in a reducing atmosphere to a fine powder with particle diameters of about 45

μm. These are blended in an acrylic resin solution or dispersion to obtain a paint with a specific resistance of 3×10^{-4} ohm/cm.[42]

Electroconductive coatings are developed from acrylic polymers, when Sb_2O_3 doped tin oxide is ball-milled in combination with the polymers. The surface resistivity is 3×10^7 ohms per square, with a sheet demonstrating a light transmission of 73%.[47]

Electrically conductive pastes for adhesives and films are prepared from 90% nickel and 10% copper. They are kneaded into a solution of ethyl-cellulose in butyl carbinol. Fifty micrometer thick films were used on laminated capacitors.[48]

Conductive pastes for silk screening are prepared from powdered metals dispersed in a solution of ethyl cellulose in a terpineol solvent. Many metallics are cited, though attention is directed to neodymium (0.6), strontium (0.4), and CoO_3, which has a low volume resistivity of 1×10^{-3} ohm/cm. Glass frits may also be added.[49]

Olefin copolymers are blended with conductive metal powders, and extruded to form sheets (0.7 mm) suitable for EMI shielding. Ethylene-propylene copolymer/metal filled sheet was reported to have a volume resistivity of 3×10^6 ohm/cm. Elongations of over 800% were possible with this material.[50]

2. Polymer plus metal flakes, metal filaments, and powders. The volume resistivities of injection molded ABS and polycarbonate with different loadings of nickel coated graphite fibers (NCG) were reported. Fibers were 7.8 μm in diameter with a 0.4 μm Ni coating. See Table 4-4[60], p. 97.

Brass fibers with a range of diameters of 5 to 50 μm and an aspect ratio greater than 5.5 were blended with ABS polymers to achieve an electrically conductive fiber mix. Horizontal churning and vertical agitation of the mix markedly influenced the volume resistivity of the composite mixture.[56]

A process is described for the addition of chopped metal fibers into the middle zone of an extruder screw feeding ABS thermoplastic. Nickel coating was 5 μm on a 25 μm aluminum fiber cut to a length of 2 mm. Forty parts by weight of Ni coated Al were added to ABS.[45]

Electroconductive film composites used for electromagnetic shielding are prepared from a mixture of fusible nonconductive thermoplastic fibers (20 μm diameter and 20 mm long) of polyethylene to which are added to 10% volume dispersion of stainless steel fibers (8 μm diameter and 20 mm long). Thin (0.1 mm) sheets formed at 180°C possess a volume resistivity of 80 ohms/cm.[46]

Conductive thermoplastics having high EMI shielding effectiveness are prepared from polymers such as poly (1,4 butylene terephthalate), polyamide nylon 6-6, poly (ethylene terephthalate), aromatic polycarbonates, and others. Composites are prepared by the synergistic combination of metal flakes (20 to 50% by weight) and metal or metal coated fibers (2 to 12% by weight).[41]

One patent for conductive plastic places emphasis upon particles of metal powder having an average particle size of less than one-third of the plastic particles. Polyvinyl chloride and polystyrene are mentioned in connection with flake silver and flake nickel-copper alloys.[58]

3. Polymers plus metal powders and carbon black. A patent reference discloses a directly electroplateable polypropylene and copolymer of a low polarity rubber. It was indicated that carbon black was the conductive filler. Carbon black in combination with metal powders is also reviewed.[62]

Acheson utilized combinations of conductive particles, C-Ag-Fe, distributed in a thermosetting binder in the preparation of conductive coatings.[63]

An electric current collector for a small battery is prepared by forming a conductive resin mixture of polypropylene (100 pbw) and carbon black (40 pbw) extruded into a 100 μm film. This conductive film is roll pressed on a 30 μm aluminum foil at 90°C and 80 kg/cm^2.[54]

A noteworthy development in metal polymer composites has appeared in electrical communication devices referred to as PTCs. The composites consist of conductive polymers exhibiting a steeply sloped positive temperature coefficient of resistance (Figure 4–8). The shape of the curve, and the sharp rise of the temperature coefficient to a peak, is typical of the semiconductors described earlier in this chapter. The composites consist of a polymer filled with carbon black powder and a metal powder, often gold, silver, or tin, which do not oxidize at the processing temperatures. High density polyethylene is the preferred polymer discussed in Reference 71. The conductive polymers dem-

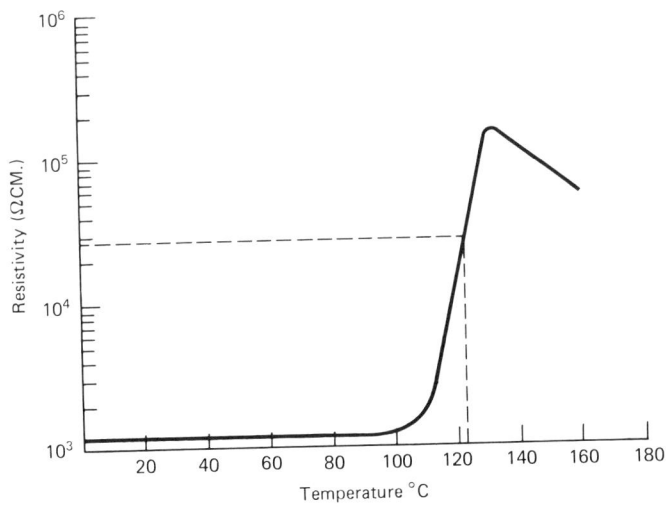

Figure 4–8. Positive temperature coefficient of resistance is a feature of this polymer/metal device.[71]

onstrating a PTC effect will function in electrical circuits, sensing ambient temperatures. They serve as current limiting devices for specific temperatures, obviating the need for thermostats.

Circuit protection devices also use conductive polymers of PTC elements, which will provide protection against sudden increases in electric current to high levels. Line circuits of telephone systems are so equipped.[72] Polyethylene and carbon black are preferred in this example, usually with a resistivity less than 10 ohm/cm. Aluminum trihydrate additives reduce the susceptibility of the PTC compositions to form carbonaceous conductive paths as a result of arcing.

4. Miscellaneous. Nickel-coated wollastonite is kneaded with polyoxymethylene polymer to prepare an electro-conductive composite. Volume resistivity of 11 ohm/cm has been obtained. It was claimed that better results were realized in those systems using nickel coated mica particles.[43]

Prevention of accumulation of long-life space-charges on D.C. power cables was accomplished by the addition of 0.2 to 5.5% carbon black to polyethylene insulation. The carbon black had particle diameters of 10 to 100 nm.[51]

Electrically conductive fabrics comprised of polyester monofilaments are prepared by electroless nickel coating with a surface layer of 0.38 μm. The 100–300 mesh fabric treated in this manner demonstrated a surface resistivity of 0.4 ohms per square.[52]

Techniques for improving the adhesion of metals to organic substrates are disclosed in an IBM patent. The subject is of interest in the study of electrical conductivity of metal/polymer composites as well as the study of metal coatings of electronic circuits. The organic substrate is irradiated with low energy particles or reactive ions (such as O or N) in the range of 50 to 2000 ev or photons of 0.2 to 500 ev (electron volts) until the surface is altered to a depth of 1 to 30 nm. Metals are deposited at a rate of 0.1 to 10 nm per second. Adhesion between copper and polyimides are improved by a factor of at least 10 over untreated polyimides.[53]

Several patent documents discuss the manufacture of niobium-tin (reinforced copper) conductors, which are used in superconductors and high energy particle accelerators. The fine Nb_3Sn conductive filaments reinforce the copper conductors. Into copper billets with longitudinal bores are imbedded Nb_3Sn strands with diffusion barriers. Although these are not examples of metal/polymer composites, increasing study of these reinforced electrical conductors will require insulative polymer or ceramic sheaths.[55] Lang has described the complex composite technology for fabricating superconductors of NbTi and Nb_3Sn which are sheathed in copper by an extrusion process. Nb_3Sn appears to be preferred because of its smaller diameter wires (3 to 5 μm) and higher temperature of superconductivity (18°K).[44] This is a striking example of an all metal composite.

The volume resistivities of injection molded ABS and polycarbonate with different loadings of nickel coated graphite fibers (NCG) have been reported. Fibers were 7.8 μm in diameter with a 0.4 μm Ni coating.[60]

P.V.C. copolymer fibers were treated on their surfaces with vacuum metallized copper. This procedure fulfilled requirements for a composite electrical conductor. Sealing with an overcoat of polymer would be a desirable measure.[61]

The aging qualities of electrically conductive polymers and of metal powder filled polymer systems must be closely scrutinized for their stability and permanence. New untried products are being produced and the prudent manufacturer will subject them to accelerated tests at high temperatures and environmental extremes. Sometimes the aging characteristics produce unexpected results, as reported by Delollis.[70] Migration of silver into the gold layer of the substrate at temperatures above 150°C, produces a rapid decrease of adhesive bond strength.

CARBON BLACKS

For many years, finely divided carbon black has been a valuable addition to electrically conductive polymers, including rubber compounds, being used to enhance their conductivity. Many patent references refer to carbon black and its ability to improve and stabilize conductivity, and the protection it offers against ultraviolet light and thermal degradation. As a reinforcement for rubber compounds, carbon black has special significance. Many grades are available from the combustion of hydrocarbon feedstocks.

Acetylene blacks have been attractive to compounders desiring high electrical conductivity. X-ray spectrographic analysis shows acetylene black to have, in part, a graphitic structure with a lace-like acicular or fibrous aggregate of carbon particles. About 70% of the particles occur in the 25 to 60 nanometer size range.[64]

More recent patents by competitive sources indicate that their carbon blacks

Table 4-4. Effect of Nickel Coated Graphite Fibers on Resistivity of ABS and Polycarbonate[60]

Percent fiber (NCG) loading by weight	0%	8.4%	10.7%	19.5%	29.3%
Injection molded ABS resistivity (ohm/cm)	3.5×10^{15}	2.2×10^5	2.1×10	2.9	1.1
Percent fiber loading by weight	0%	7.5%	15%	22.5%	30%
Injection molded polycarbonate resistivity (ohm/cm)	10^{17}	8.1×10^{14}	4.9×10^5	9.2×10^4	8.9×10^3

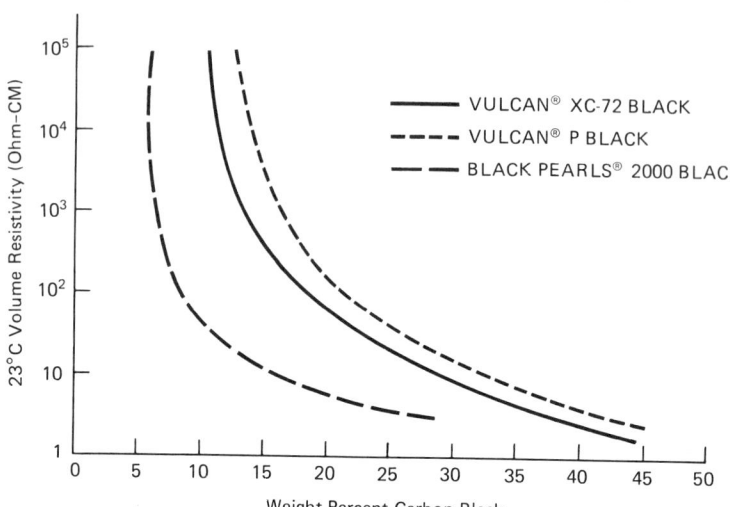

Figure 4-9. The effect of carbon black on volume resistivity. Courtesy of Cabot Corp.

and the Shawinigan carbon black noted in the preceding paragraph have particle sizes falling in the range of 250 to 500 angstroms.[65] "Vulcan" XC-72 carbon black also appears to fall in this range of good electrically conductive carbon black. "Ketjen" carbon black is also well recommended. The amount of carbon black used in the formulation of electrically conductive polymers is limited because of the contributions of metallic conductive fillers or metallic fine filaments. However, the finer particle size of acetylene blacks provides better continuity of electrical contact within the body of the material. As will be observed, graphite fibers also have been used in formulations of metal/polymer composites though the composite fibers are usually about 7 mm in diameter.

Galli has presented a good summary of the carbon blacks used by the plastics industry.[66] Furnace blacks produced by thermal decomposition of oil feedstocks, and channel blacks produced by natural gas flames on water-cooled channel iron have prominent roles in materials used by the plastics and the rubber industry. The rubber industry is, by far, the largest consumer of carbon blacks.

The Litant patent describes the use of 1 to 5% (weight) of unwoven carbon filaments in the molding compound to render the material electrically conductive.[67] Resistivities of 2.4 to 25.0 ohm/cm were reported for polyesters and epoxy compounds. Orthonitroanisole was noted as a polar plasticizer beneficial to the material's properties. Nickel coated carbon fibers offer special advantages by improving conductivity. A Sony patent reports the advantages of conductive carbon fibers in their polyester resin system.[68]

Cabot Corporation has developed data on the loading of three carbon blacks on the volume resistivity of ethylene vinyl acetate, a thermoplastic. This data is shown in Figure 4-9.[69] Carbon blacks are colloid size particles which require a substantial amount of shear to disperse properly in a plastic. As mixing time increases, conductivity increases as a result of improved dispersion. It is quite likely that the presence of metal powders (which are coarser in size) will facilitate the blending of metal powder, carbon black, and polymer into the composite system.

REFERENCES

1. Cowan, D. and Wigul, F., "The Organic Solid State," *C and E News*, 64:28, (July 21, 1986).
2. Edelson, E., "Polymers That Conduct Electricity," *Mosaic*, 19:14, (Spring 1988) and 14(2):41, (Spring 1983).
3. Maugh, T., *Science*, 222:606, (November 11, 1983).
4. Anon., *Chemical Week*, p. 56, (March 9, 1968).
5. Rembaum, A., CA78-43 986d (February 26, 1973).
6. Ferrara, J., et. al., *Applied Spectroscopy*, 41:1377, (August 1987).
7. Heeger, A., *Studies of Charge Transfer Process*, Cal Tech Conference, (November 1984).
8. Duke, C. (Xerox Corp.), *Synthetic Metals*, 21:5, (1987).
9. MacDiarmid, A., *Synthetic Metals*, 21:79, (1987).
10. Anon., *C and E News*, p. 29, (April 19, 1982).
11. Kusy, R., *Metal-Filled Polymers*, p. 51, New York: M. Dekker, (1986).
12. Skotheim, T. A., *Handbook of Conducting Polymers*, Vols. 1 and 2, New York: M. Dekker, (February 1986).
13. Delmonte, J., *Metal Filled Plastics*, New York: Reinhold Publishing Co., (1961).
14. Seymour, R. B., *Conductive Polymers*, New York: Plenum Press, (1981).
15. Baughman, R. N., et al., *"Conducting Polymers - Synthesis and Properties"* - A.C.S. Polymer Preprints, 23:130, (March 1982).
16. Jenekhe, S. A. (Honeywell), U.S. 4,717,762, (January 5, 1988).
17. Jenekhe, J., U.S. 4,631,323, (December 23, 1986).
18. Murakami, M. and Yoshi, S., *Synthetic Metals* (18), p. 509, (1987).
19. Klavetter, F. and Grubbs, R., *Polymer Material, Science and Engineering*, A.C.S., 58:855, (1988) and *C and E News*, 66:23, (October 31, 1988).
20. Roberts, W., et al., 4,604,427 (Grace Co.), (August 5, 1986).
21. MacInnes, D., *Polymer Materials, Science and Engineering*, A.C.S., 58:851, (1988).
22. Papir, Y. to Chevron Research Co., U.S. 4,519,937, (May 28, 1985).
23. Delannoy, P., et al., *Enhancing Conductivity of Doped Polyacetylyene*, U.S. 4,502,981, Allied Signal Company, (March 2, 1985).
24. Thakur, M., *Macromolecules*, 21:661, (1988).
25. Naarmann, and Theophilou, P., *New Process for Polyacetylene Films*, Synthetic Metals, 22:1, (1987).
26. Baughman, R. and Shacklette, L. *Synthetic Metals*, (17), p. 1,733, (1987).

27. CA 106(20), 166 991f and 167066u, (1987).
28. Heeger, A., et al., U.S. 4,222,903, (September 16, 1980).
29. Feldstein, N. (RCA Corp.) U.S. 3697319, (October 10, 1972).
30. Osterholm, J., et al., *Studies of Polythiophene,* Synthetic Metals, 18:169, (1987).
31. Berlin, A., et al. *Synthetic Metals,* 18:157, (1987).
32. Ramos, A., et al., *SPE Tech. Paper,* 23:393, (1977).
33. Takakura, M. et al., (Nissan Chemical), U.S. 4,666,742, (May 19, 1987).
34. Porta, G. and Taylor, L., *Jl. of Material Research,* 3(2):211, (March/April 1988).
35. Schock, F. and Partco, J., *Conductive Polyphenylene Sulphide,* ANTEC, 44th Annual Meeting, S.P.E., p. 273, (1986).
36. Wada, T., et al., *Synthetic Metals,* 18:585, (1987).
37. Lagues, M. and Sauterey, C., *Jl. of Physical Chemistry,* 84:3,503, (1980).
38. Bridge, B., *Jl. of Material Science Letter,* 7:663, (June 1988).
39. Bigg, D., "Electrical Properties of Metal Filled Polymer Composites," In *Metal-Filled Polymers,* Bhattacharya, S.K. (Ed.), New York: M. Dekker, (1986).
40. Milewski, J., *Ind. Eng. Chem. Prod. Res. Dev.,* 17:363, (1978).
41. Liu, N. and VanderMeer, R. to General Electric, *Electric Conductive Thermoplastics Composites,* U.S. 4,566,990, (January 28, 1987), and Eur. P. 185,783, (July 2, 1986).
42. C.A. 104(20), 170208g, Fukuda Metal Foil and Powder Co., J.P. 60,226,570, (November 11, 1985).
43. C.A. 107-145725c, Kuramoto, K., J.P. 62,29-007, (February 7, 1987).
44. Lang, G., et al., *Extrusion,* West Germany: D. G. Metallkunde, (1981).
45. C.A. 105-154207w, to Kanebo Ltd., J.P. 61,141,762, (June 28, 1986).
46. C.A. 105(18) 154371v, Nippon Seisen Co. J.P. 61,127,198, (June 14, 1986).
47. C.A. 105(18) 154806r, Seisuki Chemical, (1986).
48. C.A. 106-94529b, to Taiyo Jouden Co., J.P. 61,121,204, (June 9, 1986).
49. C.A. 106(14) 112171v April 1987, To Toshiba Corp., J.P. 61,107,605, (May 26, 1986).
50. C.A. 105(14) 103038e, Nippon Petrochemical, J.P. 1,236,841, (October 22, 1986).
51. C.A. 106(14) 103487a, J.P. 61,253,713, (November 11, 1986).
52. C.A. 106(14) 103740c, To Mitsubishi, J.P. 61,245,372, (October 31, 1986).
53. C.A. 106-106554f, April 1987, to IBM, E.P.A. 206,145, (December 30, 1986).
54. C.A. 106-159631k to Sumitomo Bakelite, J.P. 62,15,762, (January 24, 1987).
55. C.A. 106-161023g, Showa Elec. Wire Co., J.P. 61,276,952, (December 6, 1986); Furukowa Elec. Co., J.P. 61,279,661, (December 10, 1986); Showa Denko Kiki, J.P. 61,279,662, (December 10, 1986).
56. C.A. 106-167271h, Aron Kasei Co., J.P. 61,228,913, (October 13, 1986).
57. Sorensen, I., *Polymer Metal Blends,* to Allied Signal Co., U.S. 4,557,857, (December 10, 1985). Re also: Gergen, U.S. 4,088,626.
58. Coler, M. A., *Mfr. of Conductive Plastic,* U.S. 2,761,854, (September 4, 1956).
59. Roberts, W. and Schulz, L. to W. R. Grace & Company, *Electrical Conductive Polymer Blends,* U.S. 4,604,427, (August 5, 1986).
60. Janeczek, D. (Raytheon), *Conductive Nickel Coated Graphite Fibers,* ANTEC, 44th Annual Meeting of S.P.E., p. 226, (1983).
61. Delmonte, J., "Photodielectric Analysis of Materials," *Modern Plastics,* V. 24, p. 163, (February 1947).
62. Jones, M. to Hercules Corp., U.S. 4,002,595, (January 11, 1977).
63. Hunter, F. to Acheson Industries, U.S. 3,099,578, (July 3, 1963).
64. Kaufman, C. and Hall, R. (Shawinigan Chemicals), U.S. 2,453,440, (November 9, 1948).
65. Giet, C. (Products Chimiques, France), U.S. 4,279,880, (July 21, 1981).
66. Galli, E., *Plastic Compounding,* p. 22, (March/April 1982).

67. Litant, I. to Avco, U.S. 3,406,126, (October 15, 1968).
68. Sony Corp., U.S. 4,229,328, (October 21, 1980).
69. Cabot Corp., *Conductive Carbon Black,* Report S-39, (1988).
70. Delollis, N. (Sandia), *National SAMPE Symposium,* Albuquerque, p. 565, (1976).
71. Meyer, J. V. Texas Instruments Inc., U.S. 3,673,121, (June 27, 1972).
72. Middelman, J. M. and Doljack, F., Raychem Corp., U.S. 4,475,138, (October 2, 1984).
73. Liao, S., *Microwave Devices and Circuits,* Englewood Cliffs, NJ: Prentice Hall Publishing Co., (1980).
74. Elsenbaumer, R. L., et al. (Allied), "Stablility of Doped Conducting Polymers," *A.C.S. Polymers,* Vol. 56, p. 54, (1987).

5

PLASTICS COATED METALS AND METAL COATED PLASTICS

Coatings have long been recognized as a major product of the resin industry. Until the twentieth century naturally occurring resins had been extensively used as impregnants and surface coatings for the enhancement of other materials. With the advent of synthetic resins in this century, many new formulations have appeared and have been utilized as coatings. As decorative vehicles for colored pigments and dyes, the coatings provide attractive surfaces for an untold number of materials. Functional needs have increased dramatically in the past fifty years, and aside from their proven decorative aspects and weather resistance, coatings have contributed to structural, optical, and electrical requirements. The increasing technical demands on coatings, justify recognition of the coated products as composites. Hence, I have referred to many of the new coatings on metals or ceramics as polymer/metal composites because of the synergism of the combinations in achieving advanced technical functions.

The differences between coatings of earlier conventional lacquers and oil-based varnishes, and the newer technical coatings will be apparent from the improved durability and heat resistance of the newer coatings. The new resin binders will be considered here as will the new techniques for the application of both metals and polymers as coatings. Techniques for improved surface preparation to receive coatings will also be discussed. The topics to be treated are outlined below.

- Polymer coatings may be reinforced with glass fibers, polyaramid fibers, graphite fibers, metal wire screens, polyethylene fibers, polypropylene fibers, polyester fibers, etc., as the coatings are applied to metal substrates. Liquid polymer systems, in the absence of solvents, are generally used.
- Non-solvent mastic coatings may require heating for application. They are generally prepared from thermoplastic materials.
- Plastisols of thermoplastics in high boiling point plasticizers are an important class of metal coatings.
- The more common thermoplastics from acrylic resins, cellulose esters, and polyvinyl resins, for example, are found in myriads of applications, and

are applied to materials such as metal substrates, wood, and concrete. Partial solvent coatings from chemically resistant and solvent resistant thermoplastics, such as PTFE and PVF, require high temperature for dispersing in special solvents.
- Emulsions of thermoplastics in water support the trends to minimize the use of solvents in coatings.
- Powder coatings have become popular as a means for deposition of thick layers of plastics on electronic and magnetic devices.
- Thermal spraying of metal coatings on non-metallics.
- Electroplating of metal ions on plastics substrates requires considerable sophistication, some of the techniques include the development of conductive surfaces using electroless copper or nickel.
- Emulsions of polymers are used in the electrophoretic coating of metals by plastics. These emulsions have been widely used as primers on metal surfaces.
- Metal coatings on polymer surfaces formed by cathode sputtering and vacuum evaporation techniques, are used extensively on polymer films.
- Ablative and intumescent coatings are applied to metallic as well as non-metallic surfaces. These coatings form unique composites. They offer fire retardent properties and are useful as thermal barriers on spacecraft.

The above classes of technical coatings of polymers on metals, and metals on polymer surfaces, establish the criteria for referring to them as examples of composite materials. Thickness of film is not an adequate basis for identifying the composite. For example, the processes of metallic ion implantation discussed in an earlier chapter have established a need for composite classification. The functional objectives of attaining superior performance from two or more material combinations are the bases for composites.

In general, the conventional and well established solvent base lacquers, resin and rubber lattices, and oil soluble varnishes form coatings on the evaporation of the volatiles; the formation of the coatings are further aided, in some instances, by heat. Such coatings enjoy a very large worldwide market as decorative and protective applications to wood, concrete and metal structures. The newer "powder" coatings and emulsion coatings are taking over from liquid paints which are based on organic solvents. It is estimated that over 300,000 tons of volatile organics are released by paints into the atmosphere each year.[1]

The newer technical coatings listed above usually form films by mechanisms other than solvent evaporation. This chapter will provide information on new conceptual approaches for utilizing coatings as composites, particularly instances when metals function as the coating material rather than the polymer, which serves as the substrate. This book does not offer an in-depth evaluation of the functional roles of coating materials; such studies have been presented by others. It does however focus upon some unique interactions of metals and polymers. If there is any bias in selecting examples, it is directed toward organizations which are recognized for their leadership in these endeavors, and engineers, chemists, and scientists who have established their reputations in these fields. It must be recognized that other inorganics and ceramics, in particular, are used in many of the composite examples which are reviewed.

REINFORCED COATINGS ON METALS

Reinforced coatings for metal structures make use of fibrous reinforcements imbedded in a polymer coating matrix which is distributed over the metal surfaces. The fibers may be sprayed on the polymer coatings or sprayed simultaneously from one nozzle while the liquid resin system (free of solvents) is sprayed from another nozzle. Fibers may also be mixed into a liquid resin base before application, and trowelled in place when a thick consistency has been achieved. Thin veils of fiber or woven fabrics are used in reinforced coatings which are adhesively bonded and impregnated on to the metal surface. The most often used materials are glass fibers and woven glass fabrics, although synthetic resin fibers such as polyaramid, polypropylene, polyethylene, and graphite fibers are also used for important applications. The field of reinforced plastics is a major industry although the use of resin impregnated fibers on metal surfaces has not been widely practiced. The successful application of reinforced fiber structure in the aerospace and automotive industries, has fueled efforts to apply the same materials to metals. For these efforts to be successful, there must be consideration of the cost effective factors and objective analysis of the benefits to be derived from a fiber reinforced polymer coating on a metal substrate, as opposed to a non-reinforced polymer coating.

As an example consider a container or pressure vessel, whose contents require a chemically resistant liner of a metal such as tantalum. The container may be fabricated entirely of stainless steel or tantalum with adequate wall thickness to withstand physical stresses. The composite alternative requires that we fabricate the container of a thin interior shell of the desired metal, and apply to the exterior wall some woven glass fabrics or tapes impregnated with an epoxy resin system which is capable of being cured "in situ." The composite structure of such a cylinder would have an impervious chemically resistant interior of stainless steel or tantalum, which would be reinforced by an exterior epoxy-glass coating, which is designed to absorb the stresses. In fact, glass fiber pressure containers are extensively used in industry to contain bottled gases under pressure. An extension of this effort would introduce an inner metal liner for added strength and impermeability, if required. The metal/polymer composite would find application in special circumstances, when the polymer laminate does not alone have the required chemical resistance.

In the example described in Reference 2 a thin (0.9 mm) titanium shell was fabricated into a 51 liter container. To withstand a pressure of 300 bars, the container was wound over with 9 layers of an epoxy/graphite fiber composite. Higher breaking pressure was achieved when compared to an equivalent "Kevlar" polyamide design.

Another example in which a reinforced coating serves an exemplary function is in the maintenance and repair of sheet metal structures. Damaged galvanized steel air ventilation conduits, noncritical aluminum structural components on aircraft and boats, and sheet metal automobile fenders can have cracks repaired by the application of glass or polyaramid fabrics which have been impregnated with an epoxy resin. The low shrinkage and good adhesion of the epoxy resin system make it a good candidate for this reinforced coating application. Not only will the parts be suitable for continued service, but they will, in many structures, offer qualities superior to the original sheet metal part. Sound dampening, thermal and electrical insulation, and impact resistance would be added measures of protection. In fact, the author's first experience with reinforced coatings systems occurred when applying a layer of glass fabric and epoxy-anhydride resin to part of the surface on an experimental V-2 rocket in the late 1940s. The rocket was recovered after the test flight with the experimental reinforced coating "patch" in good condition.

Reinforced plastic coatings were reported on concrete and metals in the 1940s.[3] Solvent free polyester, phenolic, and epoxy resin systems were available then for the addition of glass fibers, which added strength and durability. Glass fabrics were first available in the 1930s. These fabrics took advantage of solvent free epoxy and polyester resins for technically advanced reinforced coating in the late 1940s. The lower shrinkage and better adhesion of the epoxy resin system prevailed over other polymers. Today, preimpregnated glass fabrics, polyaramid fabrics, and graphite fabrics may be purchased from prepreg manufacturers and suppliers of potential reinforced coatings. Newer high temperature polymer systems and improved surface preparatory measures are available to sustain this activity on metal/polymer composites. Laminating techniques, such as filament winding and extrusion, reviewed in Chapter 6, should be considered for large scale production of metal/polymer composites. From a coatings viewpoint, there are advantages to be considered in the use of adhesive, reinforced fabrics and tapes of high strength fibers. Such products build up a substantially thicker layer, much more rapidly than multiple sprayed coatings of a lacquer or liquid polymer (which would tend to sag during cure).

NON-SOLVENT MASTIC COATINGS

Non-solvent coatings require the use of liquid polymers and catalysts, or curing agents, which make possible their conversion to a solid film. As noted earlier, the reinforcement of fibers will facilitate application of thick layers. An objective is to obviate the use of multiple coatings to build a thick coating (of about

1 mm or more) on a metal surface. The principle is understandable, but as one considers the research and experience developed for automobile body exteriors, the complexity of the task becomes obvious. Consideration of primers for adhesion, pigmented intermediate and finish coats, which are sprayed on the bodies moving on an assembly line at a prescribed rate and temperature cycle, lead to good results. The commercial systems for automobile exteriors involve solvent systems and electrophoretic coatings to attain beauty and durability of the metal bodies. Such coatings could be called polymer/metal composites, but in this section the lesser known non-solvent coatings will be described because of specific functional demands.

Underbody, *thick* mastic coatings to protect the automobile body from dirt and the chemicals of snow removal would be non-solvent coatings. Bituminous-based coatings can be liquified by heating, developing a sufficiently low viscosity for spraying, and allowing a rapid build-up of layers on a metal substrate. The thick semi-resilient, wear-resistant coatings are useful components on a polymer/metal composite. These same concepts apply to readily meltable and sprayable thermoplastics. Additives, such as rosin esters and cumarone-indene resin would be useful, fluidifying additions to such formulations. Fillers could include metal powders and fibers, and inorganic substances to enhance the properties of these mastic-like coatings which are applied to metals. The presence of volatile substances in the applied coatings contributes to instability in the form of shrinkage, cracking, and unsatisfactory moisture resistance. Because of such factors thick mastic-like barriers for metals, whether as seals or crack fillers on metals, form the basis of non-solvent coatings. They are not necessarily applied as smooth, colorful surfaces (as found on visible exterior automotive finishes) but for protective purposes.

There has been significant activity displayed in the issuing of recent Japanese and United States patents for polymer resin systems, and for fiber reinforcements on the underside of automobile chassis. In addition to providing protection against road surface hazards, polymer/metal composite systems are effective in reducing noise from the underside of vehicles. The systems must be designed with good adhesion between the metal and the polymer and/or between mastic, and exhibit resistance to mechanical vibration under cold weather conditions. Fiber reinforcements contained within these thick underbody coatings provide additional reliability. In Chapter 3 the roles of metal powders and polymers were described as trowellable coatings for automobile bodies, ship maintenance, and storage tanks. With additional fiber reinforcements these products are capable of withstanding additional stresses. Glass, polyaramid, and graphite fibers in a woven form extend the protective properties.

There are other techniques which may be practiced for the exterior coating of continuous metal components such as tubes, rods, I-beams, and other structural elements:

- Thermoplastic tubing is available in a heat shrinkable form. The material is applied over a metal rod or tube. The introduction of heat causes the thermoplastic to shrink and seize tightly about the profile.
- Woven sleevings of glass fiber may be slipped over a metal profile. After being pulled taut, the fibers draw tightly about the metal and serve as a base upon which resin impregnants such as epoxy, polyurethane, or polyester may be deposited.
- The complex weaving of fibers may be coordinated with the creation of unusual profiles to insure complete coverage of a wood or metal profile (Figure 5-1). A resin coating may then be applied to impregnate the fibers. This technique offers the advantage of continuous lengths of shaped structural fibers. Great structural benefits are obtained in this manner. These benefits are superior to those obtained by the random inclusion of chopped fibers into the polymer coating. In Chapter 6 a section on filament winding discusses these matters.

Designers of space stations and lunar stations are aware of the advantages of metal cores, which use resistant coatings of organic polymers, for stiffness and strength. Radiation in space affects the properties of the construction ma-

Figure 5-1. Braided glass fiber coatings over metal and plastics products. Courtesy of Owens-Corning Fiberglass Corp.

108 METAL/POLYMER COMPOSITES

terials, and metal/polymer composites offer prospects for the solution to the problem of the degradation of the materials.

PLASTISOLS

Polyvinyl resins, particularly polyvinyl chloride (PVC), are dispersed as fine powder in nonvolatile liquids known as plasticizers. These dispersions are called plastisols, and depending upon the proportions and type of plasticizers used, they will vary in viscosity from a motor oil consistency to a putty-like consistency. There are other products, referred to an organosols, which use volatile components to disperse the vinyl powders for maximum fluidity. Flexibility plasticizers are added to these products. The organosols are adapted to spray techniques. At the time of spraying the volatile solvents evaporate. Plastisols are usually used to coat metal products and the two materials have been combined to form many thousands of composites for consumer products.

The presence of large amounts of plasticizer such as dibutyl phthalate, tricresyl phosphate, dioctyl phthalate, and many others, contributes to the variability of physical properties. The more flexible materials, which are used as coatings, will have properties such as those noted in Table 5-1.

More specific information may be obtained from Figure 5-2 which shows the relationship between the physical properties and the formulation of GEON 121, one of the most widely used polyvinyl resins for plastisols.

Application of plastisols to irregularly shaped metal objects such as electroplating racks, refrigerator shelves, clothes hangers, dish-drain baskets, tool handles, etc., is achieved by dip-coating. Two processes are available: hot dipping and cold dipping. In the hot dipping process, the metal product is cleaned and degreased, coated with an adhesive film if desired, and moved into an oven to raise the temperature of the metal prior to immersion in the plastisol. The temperature of the metal object gels the polymer deposited on the surface. Further oven heating completes the fusion of the plastisol coating. Coatings up to 4 mm thick are typical deposits for a wide selection of colored vinyl coatings on wire racks for dishes and appliances. Typical products are

Table 5-1. Typical Physical Properties of Polyvinyl Chloride Plastisols

Tensile strengths	1.5 (MPa) to 250 (MPa)
Elongation percent	500 to 100
Shore A hardness	20 to 95
Volume resistivity (ohm-cm)	10^{14} to 10^{15}
Dielectric strength (3 mils thick)	400 volts/mil (16 volts/mil)
Service temperature (continuous)	60°C to 90°C
Arc resistance to flashover	Good

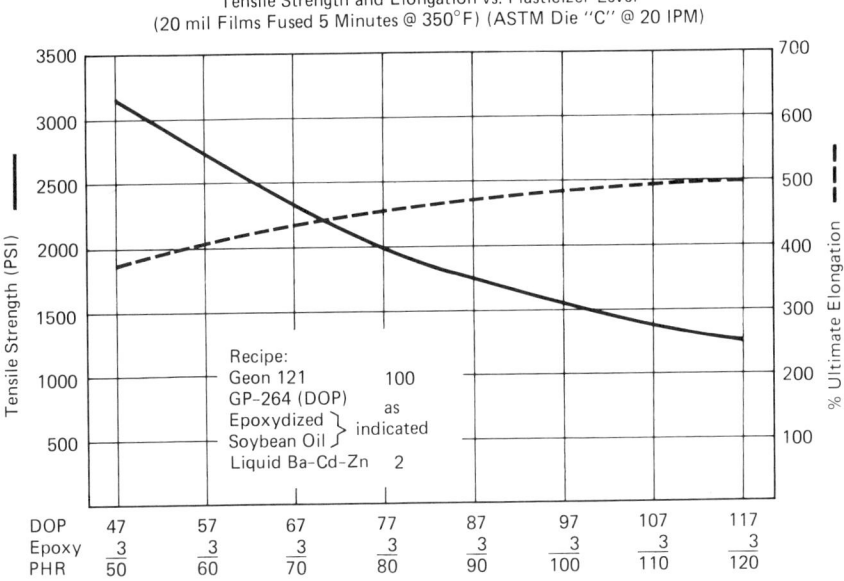

Figure 5-2. The relationship between plasticizer level of DOP (Dioctyl Phthalate) and "Geon" 121 and physical properties of vinyl plastisol.[4] Courtesy of B.F. Goodrich Co.

shown in Figure 5-3. The rubber-like feel of the vinyl plastisols, and their moderate prices, have made them popular in the consumer home appliance field. They are limited in their service temperature and their adhesive qualities to metal.

In the cold metal dipping process of application, conveyor lines transfer the objects to the vinyl plastisol bath without preheating. After immersion, which typically forms coatings of 0.1 to 1 mm, the coatings are baked to complete plasticization and fusion. Particular care is taken, by removing parts at a controlled rate from the plastisol bath, to avoid dripping or sagging from the metal objects being coated.[4]

When large sheets of metal are to be coated, semi-pastes of polyvinyl resins may be used and applied by roller and knife coaters. Viscosity characteristics of the vinyl compound, speed of coating, clearance between the web and knife or roll, and angle of the knife are all factors to be considered. Heavy fabrics and papers may also be treated in this manner. Plastisols may also be used to fill cavities formed by metal shapes and hence establish composites with metal exteriors.

Temperature resistance of these coatings is limited because of the high plas-

110 METAL/POLYMER COMPOSITES

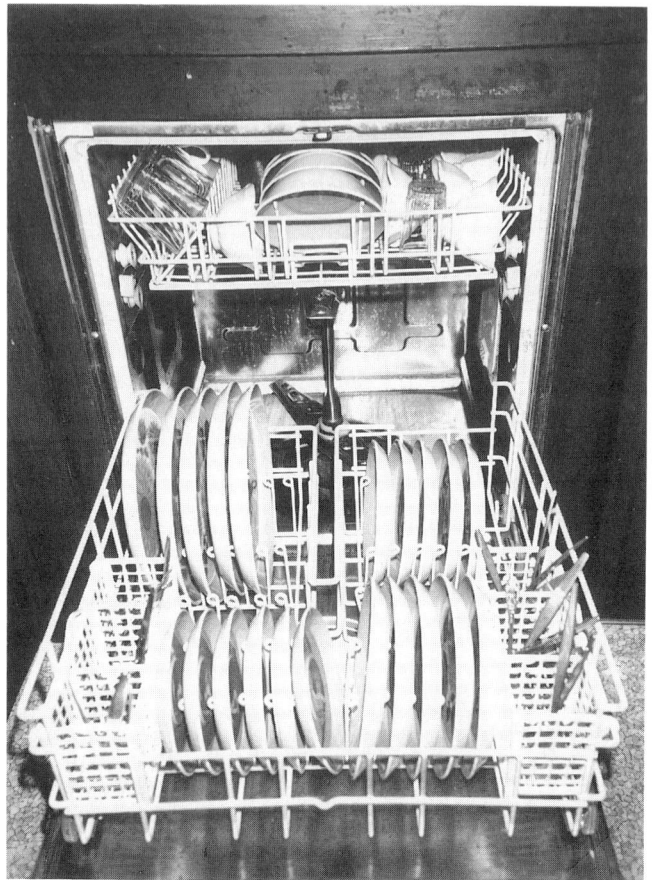

Figure 5-3. Plastisol coated wire rack for dishwasher.

ticizer content, but the process does achieve a substantial deposition of flexibilized polymer upon the metal. Adhesion to the metal, while not high, is adequate as the metal parts are usually encapsulated in the plastic to form the composite.

Vinyl plastisol processing necessitates heat stabilizers for some of the formulations developed from polyvinyl resin systems, particularly when steel and zinc parts are involved. Thermal aging of the PVC/metal composite produces some degradation of the plasticized polymer, accompanied by discoloration.

DISPERSION COATINGS

There are some thermoplastics which are not readily soluble in conventional solvents and are not easily plasticized. They have other characteristics such as good heat resistance and good chemical resistance which establish them as candidates for special coatings on metals. Polyvinyl fluoride (PVF) and polytetrafluoroethylene (PTFE) are typical polymers which meet these requirements. Finely divided powders are dispersed in non-solvents and blended with compounding ingredients at high temperatures before application to the metal products. Further baking is necessary to remove the volatile matter which was introduced.

Polyvinyl fluoride (PVF) and polyvinyl difluoride (PVDF) have particular merit because of their outstanding resistance to weathering, superior heat resistance, and good dielectric properties. PVF and PVDF films are being used in new electronic devices, where their characteristics as a piezoelectrical material, or as an electret have particular value. The product is mentioned in connection with electronic devices in the last chapter of this book. Because of its relatively high cost (as compared to PVC) it has not been used in the consumer products noted above, but it has been singled out in new advanced applications. Solubilization is not readily attained in ordinary solvents, and plasticizers are not generally utilized in developing PVF coatings.

PVF dispersion coatings are applied to preheated (200°C) aluminum metal objects. The dispersion coating bath, consisting of PVF powder in propylene carbonate, is kept at a temperature below 100°C. The viscosity of the bath lies in the range of 2500 centipoise to 20,000 centipoise. Non-sagging uniform coatings are obtained after coalescing them at 185°C for about 8 minutes. The required thickness is achieved by controlling the temperature of the metal-substrate and the duration of immersion.[5]

Glass fibers, treated with silane and titanium coupling agents, are blended with 20% PVF at high temperature in a solvent. The coating is applied to an epoxy primed galvanized steel sheet and baked at 240°C for 50 seconds. The 25 μm coating offered enhanced abrasion, weather, and corrosion resistance.[6]

Polytetrafluoroethylene (PTFE) coatings make available unique polymer/metal composites of exceptional chemical and heat resistance. Like PVF, the PTFE polymer (Teflon) is not readily dissolved at room temperatures in conventional organic solvents. However, the proliferation of Teflon lined cooking pots and pans in virtually every household in the United States, bears witness to the unique chemical resistance of the material and its value as an antisticking surface for many food products (Fig. 5-4).

Another good industrial example is a 4500 gallon tanker lined with Teflon fluorocarbon resin for transporting high purity hydrogen peroxide. H_2O_2 is used

112 METAL/POLYMER COMPOSITES

Figure 5-4. Metal cooking pans coated with "Teflon." Courtesy of E.I. DuPont de Nemours.

by the semiconductor industry as a cleaning and etching agent in the manufacture of computer chips.

PTFE copolymers have been bonded to metal and to ceramic structures without chemical or physical modifications up to 30% (wt.). Inert particles of Fe_2O_3 of 0.5 μm size are bonded at 160°C with the polymers to form a sheet 0.3 mm thick. This adhesive sheet pressed between stainless steel and PTFE shows a peel strength of 4.0 kg at 25 mm.[7]

Some organic metal complexes when dispersed in a polymer solution have the capability, after application, of forming a metal layer on the polymer film upon subsequent heating. The size of the metallic particles is very fine and at a level smaller than one micrometer. Organic metal complexes which will release fine metal particles were reviewed earlier—and it has been claimed that, unlike mixing of powdered metals into polymer solutions, the thermal release (at 150° to 200°C) of metal particles from the organic complexes achieves a uniform surface coating on the polymer. A number of organic metal complexes were listed, including metals of Group IVA, VA, VIIA and VIII of the periodic table and ligands from triphenyl phosphine, tributyl phosphites, cyclopentadiene, etc.[8]

Nonaqueous, colloidal dispersions of polysulfone polymers have been prepared for electrophoretic deposition on metals. The polymer, containing sulfone groups in the polymer chain, is blended with 20 to 37 parts by weight of a nonaqueous solvent and 0.8 to 1.2 parts by weight of a tertiary amine, and colloidally dispersed in a ketone. Good electrophoretic coating of complex metal parts are claimed.[9]

Figure 5-5 shows a circuit board containing metal, ceramic, and plastic components which has been sealed with a reversion resistant polyurethane coating, thus forming a multi-composite assembly. The reversion resistance to high humidities characterizes this coating.

PLASTIC COATING OF METAL STRUCTURES

The cladding of metal structures by plastics has been accomplished from solutions, emulsions of polymers, and the spraying of powders upon a heated substrate. The term *cladding* suggests a relatively thick coating, with distinctive functional objectives, and final products which are composites. Surface preparatory measures must be planned to remove contaminants, particularly

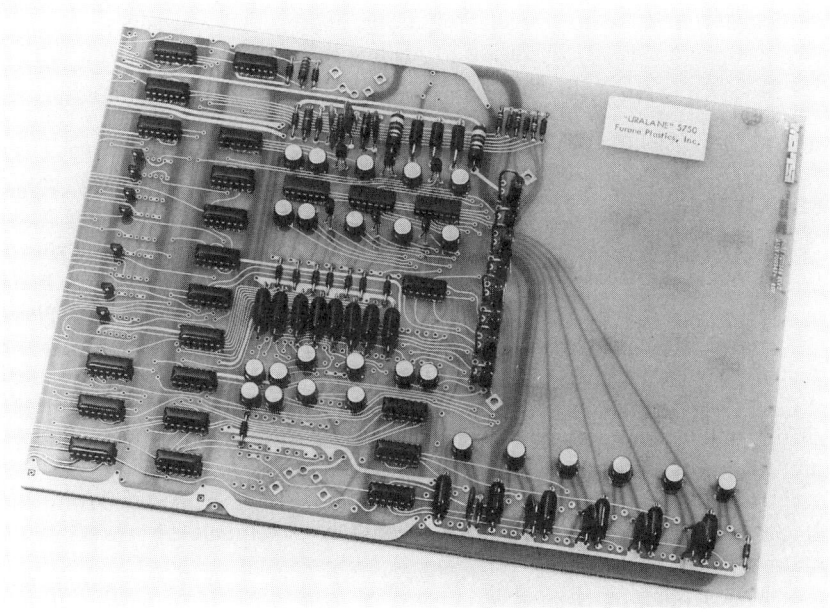

Figure 5-5. Circuit board coated with Furane's Reversion Resistant Urethane. Courtesy of Furane Division of Ciba-Geigy Corp.

petroleum derivatives, as well as corrosion products, and loose scale (as from steel). In principle, clean, moisture free surfaces, which may have a roughened profile to facilitate adhesion, are mandatory for the thickly clad structures. Examples of plastic coated metal structures are numerous but a few examples of these fortuitous composites would include:

1. The hundreds of miles of epoxy coatings on the interior of the steel oil pipelines (over 6 feet in diameter) from the Alaska oil fields, have performed satisfactorily since the early 1980s. The coatings were designed to handle the heated oil and also abrasive sludges from the oil fields.

2. Marine environments are particularly hostile to metal products, and the preventative maintenance costs of establishing corrosion resistance are significant. Naval shipboard coatings have been continuously improved over the years with the emphasis concentrating on heavily pigmented oil resin bases which offer protective shields and the ability to yield to physical stresses arising in high seas. The approval of shipboard coatings by the U.S. Navy requires long periods of field testing. Proposals for some experimental coatings have required fiber-reinforced coatings on top of the sheet metal used in a ship's structure. The writer has been involved in refurbishing a ship's propeller shaft which was built up and overlayed with glass fibers and epoxies. Overlays of glass fibers and fabrics may be accomplished upon metal structures in the field, particularly when room-temperature curing impregnants and adhesives of epoxy resins are used. Heat curing of resin systems was possible with infrared lamps.

Unique to ocean going vessels are a special class of anti-fouling coatings which are used to counteract barnacle growths which develop on metal ship hulls. If their hulls are not properly maintained, these vessels may have a serious loss of surface speed, and higher fuel requirements. Anti-fouling plastic coatings have, historically, depended upon the slow release of toxic chemicals to inhibit the growth of barnacles. After the passage of several years, as the coatings are slowly depleted of their toxic chemicals, they have to be replaced in the shipyard. Most metals and films, such as chemically resistant polytetrafluoroethylene, will harbor the barnacles on their surfaces. Hence, the need for a metal/plastic composite with chemical deactivating agents. Tributyltin oxide additives are now being studied on U.S. Naval vessels, though copper anti-fouling compounds have been used for years.[10]

3. The *glassing* of small boats became a popular practice after World War II. Small boats (both wood and metal) would have their hull surfaces reinforced and waterproofed with glass fiber sheets, and polyester or epoxy resin systems which were substantially solvent free. For more sophisticated, large volume requirements, *spray-up* techniques of chopped fibers and resin systems were used to cover large areas economically. Although these practices still prevail,

the use of all glass fiber-resin boats and laminated composites has opened new fields for development.

4. The lamination of metal sheets and foils to polymer films and sheets provides large area combinations of composites. In some instances, they represent larger size applications in developing metal/polymer composites than could be attained by plating and spraying procedures.

The application of polyester coatings to metal sheets requires careful procedures to obtain sheets with good corrosion resistance and good adhesion, even after severe forming operations. Attention is directed to the use of a polyethylene terephthalate film applied in a thickness of 10 to 30 μm.[35] The resin film is precoated with an epoxy resin adhesive having an epoxy equivalency of 400 to 4000 and a phenolic resin curing agent produced from paracresol. The adhesive is applied in the range of 1 to 2 g/m². If the epoxy polymer resin equivalency is too low, tackiness causes difficulties when an attempt is made to uncoil the polyester resin film which has been precoated with epoxy adhesive. Above 4000 epoxy equivalency, the adhesive strength for the metal sheet and polyester film is not adequate.

The steel plate which is to be coated with the adhesively coated resin film is first plated with less than 3.0 g/m² of nickel. The plate is next treated cathodically with a chromic oxide (CrO_3) electrolyte to yield a surface containing a hydrated chromic oxide (8 to 20 mg/m²). When the laminating occurs the adhesive coated resin film is bonded to the steel sheet by heated rolls (the temperature of the steel sheet is about 240° to 250°C). These details demonstrate that commercial uses for large scale application of polymer coatings on metal require careful study to select pretreatment procedures adaptable to the available processing temperatures and pressures, and to remain within economic time limits. Tests for the suitability of polymer coated steel sheets require that the materials undergo severe bending stresses and environmental exposures to affirm suitability of the composite product.

5. Ultraviolet cured coatings of polyester and epoxy polymer systems are popular on metal components designed for electronic circuits. The rapid cure possible with ultraviolet catalysts makes it possible to establish procedures for fast moving assembly lines. The ultraviolet cured polymer films are generally thin and sensitized at their top surfaces.

POWDER COATINGS

Finely divided plastics formulations have become popular in recent years as coatings for metal components and for devices used in the electronic and appliance industries. Their development was inspired, in part, by a desire to eliminate volatile solvents from the atmosphere and from the work place. Un-

116 METAL/POLYMER COMPOSITES

like the plastisols described earlier, the powder coatings are much more durable, harder in texture, more tightly adherent to metals, and aesthetically more pleasing in appearance.

Experience has demonstrated that when the plastics powder coating fuses and/or cures on the metal component, that rounding of sharp edges on fabricated metal parts will take place. This does not easily occur when solvent sprayed polymer coatings are applied because the surface tension of the liquid phase of the coating causes the coating to be pulled away from the edges of the surface. Compounding liquid coatings with fine fibrous powders and thixotropic agents, such as colloidal silica, minimizes this tendency. However, the build up properties of powder coatings over sharp edges have definite advantages in uniformity and quality. (Fig. 5-6).

In the conventional dry powder blending and coating processes, the materials are usually polymers, coloring agents, and flow control agents which are used to insure uniformity of distribution of pigments (frequently oxides of metals) and the polymer components. Prior thermal blending may be imperative. Blending would be followed by grinding, and reduction of the material to the particle size range desired for the application process. Because of their comparatively low cure shrinkage and good dielectric properties, epoxy resin systems are preferred, among the thermosetting resins, for the coating of metal

Figure 5-6. Epoxy fluid bed coatings are successfully applied to transformers and metal hardware.

components. When operating temperatures are not too demanding, polyurethane resin systems are noted for toughness and impact resistance for some classes of circuit board coatings.

The metal component or its assembly must be preheated to a high temperature so that immersion into the powder will fuse the formulations. Heavier masses and thicker sections hold the heat longer, but immersion time is short and the complete fusion and/or cure occurs in a separate oven. The powder may be maintained in a *fluid bed* where the finely divided particles are mechanically agitated and vibrated, to attain a loose, porous mass into which parts are easily inserted. Heated parts are immersed for several seconds into this *fluid bed*. An alternative procedure is to spray the compounded powder on to preheated metal components, which are moving on an assembly line into a spray chamber. More material is lost and wasted in the latter procedure than in the fluid bed chamber procedure.

The potential applications of the controlled build up of the powder coating procedures are unlimited. Base layers of pure polymers deposited on the metal component may be followed with powders which possess good thermal and electrical conductivities. Or alternatively, unique insulative qualities may be introduced and finished with a decorative or functional top coating. Controlled and enhanced properties are characteristic of the polymer/metal composites formed by powdered polymer compositions deposited upon metal components.

Lawn and patio furniture, appliances, laundry tubs, etc., are receiving durable powder coatings. The thermoplastic particles are applied with an electrostatic spray gun at 60 to 100 kilovolts. Powder particles adhere to a preheated substrate, as they melt and coalesce into a smooth durable coating (without the presence of solvents).

A technique of the Ferro Corporation applies powder coatings to a mold surface, which then becomes an integral part of the substrate as it is molded.[11] This demonstrates the versatility of powder coatings which do not fuse until they contact the hot mold surface. Powder coatings may also contain electrically conductive fillers; in one technique the epoxy surface coating was ground with graphite to yield a volume resistivity of 10^{10} to 10^{12} ohm/cm. The coating was electrostatically applied to a steel sheet and baked.[12] Superior corrosion resistance to salt water was claimed.

SPRAYED METAL COATINGS ON PLASTICS

While there is ample room for a variety of formulations among the polymer compounds intended as sprayable materials for coatings, the formulations of sprays for metals to nonmetallic surfaces are limited to highly filled fine metal powders dispersed in a polymer solution (silver and gold for example), or to metal coatings applied by arc spray devices. One of the principles is depicted in Figure 5-7, in which an arc is established between contacting wires, forming positive and negative electrodes. The consumable wire electrodes, connected to a direct current source of power, form a narrow gap and create an arc hot enough to melt high temperature metals. As the feed wires melt, the molten metal drops are atomized and propelled to the work surface by a blast of compressed air. Temperatures at the surface being coated are high and heat resistant materials are necessary. Metal powders for thermal spraying are characterized by chemical composition, particle size, and morphology. For lower temperature metal spraying on plastics, zinc spraying from wire stocks is used extensively.

In Chapter 7 attention is directed to zinc sprayed on molded plastics for shielding against electromagnetic interference (EMI). The continuity of the deposited metal coating has established the performance of good quality EMI shielding against which wire grids, vacuum metallized coatings, electro deposited coatings, and metal fiber filled molding compounds are compared.

In another recent development, a typical multi-component composite is formed during the preparation of light weight concrete block building materials. The concrete blocks are thermally spray coated with copper or bronze alloys, and the metal surfaces are overcoated with plastic materials. The porosity which would ordinarily develop on the metal surfaces is sealed with polyurethane coatings, improving the material's durability for weather resistance and its ability to withstand cold temperature cycling.

Plasma spraying can be used to melt and apply a number of materials including refractory ceramics. The plasma gun consists of a cone shaped cathode

ARC SPRAY DEVICE

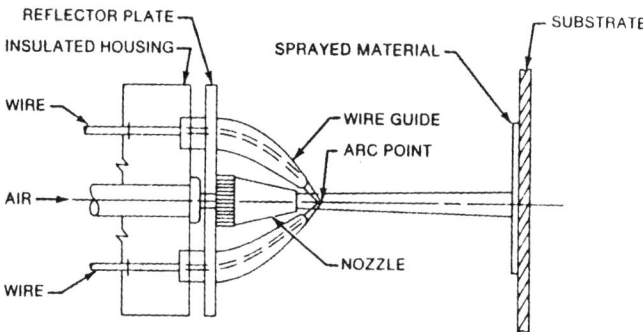

Figure 5-7. Schematic view of metal spraying on plastic substrate. Lower melting point metals are preferred for this method.

inside a cylindrical anode which extends beyond the cathode to form a nozzle. An inert gas, usually argon with some hydrogen, passes between the electrodes, where it is ionized to form a plasma.

Powdered coating materials (metals, polymers, or ceramics) are directed into the plasma arc. The gas within the arc—the arc has been initiated by a high-energy pulse of direct electric current—is transformed into a collection of ions and energetic electrons. Layers of material, 10 μm or more in thickness, may be applied by plasma coating. Earlier successes have appeared on ceramic coated metallic engine and turbine components.[37]

ELECTROPLATING METALS ON PLASTICS SURFACES

There are effective commercial procedures for electroplating plastic materials. Early developments in the preparation of mirrors by the chemical deposition of silver films on glass surfaces paved the way for rendering surfaces of non-metallic materials electrically conductive. Although such films are fragile, they can be reinforced by the electroplating of copper, nickel, and other metals in sequential steps described in this section. Parts are immersed in multiple cleaning and sensitizing baths, which furnish metallic ions that move towards the electrode, which are the pre-applied electrically conductive films formed on the plastic parts. Recently silver conductive treatments to achieve electroconductivity have been replaced by other metallic elements, applied by electroless, chemical techniques. The preplating process is followed by the electroplating process to obtain thicker, more durable metallic films. In fact, several metallic films such as copper, or nickel may appear on the final composite. Such electroplated polymer composites offer attractive materials with their mirror-like finish, and their ability to perform functions on electrical and electronic devices. These will be described in Chapters 8 and 9.

Metal plating on plastics has been practiced for more than 50 years. Acrylonitrile butadiene styrene (ABS), polypropylene, phenylene oxide based resins, and polysulfones have been among the more widely used materials. In the earlier years, silver or graphite coatings were applied to obtain electrically conductive surfaces for further electro-deposition. The ability of strong oxidizing solutions to chemically etch micropores on the plastic surface influences the preplating of the above polymers. Preplating by the electroless metal deposition process utilizes a complex tin-palladium compound to introduce active palladium sites into the micropores. Electroless metal deposition (copper or nickel) provides electrical conductivity for subsequent electroplating on the plastics.[13]

Solutions for auto-catalytic deposition of either nickel or copper are used to render plastic surfaces electrically conductive.[14] A uniform metal film 0.25 to

0.50 μm thick is deposited. Additional electro-deposited layers of either copper or nickel are applied at low current density from suitable strike baths. Metal thickness is increased to 2.5 to 4.0 μm to facilitate electro-deposition of decorative or functional coatings by conventional means. Plated plastics have altered physical properties due, partly, to the metal plate itself and, partly, to the plating process.

The preparation of substrates to receive metal by electroplating can be enhanced by irradiation with low energy particles, reactive ions (O or N ion beams), or electrons at 50–2000 electron volts (ev) or photons of 0.2 to 500 ev energy until the surface is altered to a depth of 1 to 30 nm. Copper is electro-deposited at a rate of 0.1 to 10 nm per second. In one example the substrate was polyimide and the adhesion of copper to the polyimide was increased by a factor of 10 over the untreated polyimide.[15]

Polyimide films with implanted Cu^+ and Zn^+ ions were studied recently. The implanted surface layers of the polymer were carbonized by polymer scission. Tiny islands of metal were established with high doses of Cu^+ and Zn^+ ions.[16]

A majority of electroplated plastics are finished with copper/nickel/chrome electroplate with smooth surfaces upon defect free molded parts. Mirror-like finishes are obtained. Finishes are seldom based on a single metal; usually two or more are applied with specific purposes. Thus for exterior automotive parts a layer of bright acid copper acts as a brightening and leveling layer, and as a cushion between the plastic substrate and the semi-bright nickel, absorbing differences in linear thermal expansion rates between the plastic and nickel plate.[17] The semi-bright nickel layer enhances corrosion resistance, and the bright nickel plate helps to determine color and brightness, particularly if followed by a flash of chromium plate for hardness.

Table 5-2. Range of Peel Strength Values for Electrodeposited Copper and Nickel on Typical Plastics

TYPE OF PLASTIC	PEEL STRENGTH ON 25-MM WIDE STRIPS COATING THICKNESS 35 ± 5 μm	
	Copper	Nickel
ABS	40–110	40–110
Modified polyphenylene oxide	10–50	10–50
Polypropylene	20–110	70–125
Impact resistant polystyrene	9–20	9–50
Polysulfone	20–100	20–100
Polycarbonate	10–50	10–50

*Peel strength forces in N (Newtons) measured as per ASTM 8533-85.

Electroplating for EMI shielding uses primarily electroless-plating since only a thin, uniformly thick layer is necessary. A two deposit system is generally preferred, with an electroless copper for the EMI shielding and a top layer of electroless nickel to protect the copper and to function as a paint base for decorative purposes.

A technique for the metal electroplating of polymers of poly (aryl ethers) and copolymers, which are stabilized with a polyhydroxy ether, includes considerable details on the preparation of an electroless copper bath. Compositions were designated for metal/polymer composites of plumbing ware and appliances. Adhesion tests were performed on a final copper electroplate of about 100 μm. Cycling tests from high to freezing temperatures were satisfactory.[18]

In a different approach to the application of metal plating to plastics articles, the electroplating is applied to the mold surface, which shapes the plastic article. The metal plating transfer is aided by nodular growths on the inside of the electroplate. Mechanical adhesion develops on the enveloping plastic material during molding. For some specialized molded polymer/metal composites, there may be advantages to the procedures.[19]

In earlier developments, the metallizing of a nonconductive substrate was accomplished by precoating the surface with an organic polymer/carbon black mixture containing sulphur having a volume resistivity less than 10^3 ohm/centimeters. The treated substrate is used as a cathode in a nickel, cobalt, or iron plating bath.[20] Later developments have replaced this process which was intended for elastomeric products.

A process for electroplating directly on a plateable plastic surface (pretreated to be electrically conductive) with a thin nickel-cobalt alloy strike has been described. The electro-deposited alloy is 0.1 to 0.5 μm thick and is directly coated with a nickel electrodeposit at least 0.9 μm thick.[21]

Glass fibers have been treated with traces of a titanium coupling agent and then electrolessly plated with nickel. A composition of 100 pbw of ABS resin and 35 pbw of nickel coated glass fibers yielded parts with good electromagnetic shielding.[20]

POLYMER PLATING ON CONDUCTIVE (METAL) SUBSTRATES

When an electric field is applied to a colloidal dispersion of a polymer, the colloidal particles migrate to the anode and the cathode in the same way as the ions in a metal plating bath. The process is called electrophoresis and as the colloidal particles collect at the electrode, an insulating sheath is built up. Hence the thickness deposited is restricted, usually in the range of 20 to 30 μm or less, if desired. The coating thickness achieved is very uniform, including edges, projections, and corners of metallic articles. The dispersion is usually developed in a water matrix, obviating problems relating to solvent disposal. This process was pioneered in London at Crosse and Blackwell Ltd. during the 1930s.

The interest in this technique of applying plastic coatings on metals is high and the technique is looked upon as a method for enhancing the quality of polymer/metal composites, particularly on electronic devices and semiconductors. In paragraphs below are summarized recent activities involving electrophoretic applications of polymers on metals.

Recent patents on electrophoretic depositions of polymers on metal surfaces will provide details of the procedures that are followed. Polysulfone polymers and polyether sulfone polymers are dissolved in a nonaqueous liquid such as dimethyl acetamide, dimethyl formamide, n-methyl 2 pyrrolidone, dimethyl sulfoxide, or pyridine. This polymer solution was rapidly added to a nonsolvent of acetone to form an emulsion. The volume ratio of nonsolvent to solvent is about 3.5 to 5.0, and the solids content of the emulsion is about 0.4 to 0.7% by weight. Coating on the metal anode (aluminum) was developed by the application of 300 volts dc. The counter electrodes were nickel screens separated from the anode at a distance of about 3.5 cm. The thickness of a typical polymer coating was in the range of 20 to 30 μm.

Water based coatings are applied to metal automobile bodies as a primer with the object to be painted as one electrode of the cell. When an electric current passes through the liquid, the binder moves toward the electrode, where it deposits itself, carrying pigment particles which are attached to the binder. For years the negatively charged carboxylate salts were the binders and headed for the anode, while the neutralizing ion moved in the opposite direction. Recently, positively charged polymer molecules have been used as the binder; they travel to the cathode rather than the anode in the painting bath.[1]

Surface treatments of titanium were accomplished by electrophoretically applied coatings of polymers solubilized in water by tertiary amines or quaternary ammonium salts. The titanium functioned as an electrode.[21]

Water thinned acrylic polymers have been electrodeposited upon cobalt-tungsten alloys. The factors affecting the product were related to solid content of the liquid (8%), pH value (7.5 to 8.1) and the application voltage of 40–60 volts dc.[22]

Several polymer coatings were identified for applications involving lead alloy grids used in lead/acid storage batteries. A 2- to 3-fold life extension is expected over batteries using uncoated grids. The polymer coatings included acrylic polysiloxane binder, chlorosulfonated polyethylene/epoxy blend, PTFE, copolymer of vinylidene fluoride, and hexafluoropropylene. The coatings were rendered electrically conductive by carbon black and graphite. They were applied to the lead grids as layers of 10 to 200 μm thickness.[23]

Coatings of flexible printed electronic circuit boards of polyimides (25 μm with 18 μm of copper) were electroplated in a nickel sulfamate bath with a 1 to 2 μm strike of nickel. This was followed by a gold plating of 3 μm to get good crack resistance and bonding qualities. The multiple layering of metal coats on a polymer substrate is one of the techniques used for the formation of polymer/metal composites for electronic components.[24]

Developments in conformal coatings and new plating chemistry, as applicable to metal/polymer composites on electronic circuit boards, continue to be announced at trade shows and in the technical press. Among developments relevant to the coating of circuit boards have been the items noted below:

- Parylene coatings. Parylene is a generic name for various members of a polymer series (para-xylylenes—(CH_2—C_6H_4-CH_2) by Union Carbide Corp.). They have been in use for approximately 20 years, primarily for protecting delicate microcircuitry from adverse environments. The parylene polymers are deposited from the vapor phase by a process resembling vacuum metallizing. However, whereas typical vacuum metallization takes place at very low pressures, the parylene coating (typically 10 to 15 μm thick) is applied uniformly under more moderate conditions. Vaporization of the dimer solid of paraxylylene takes place at 160°C, followed by pyrolysis of the dimer at the two methylene bonds at 680°C to yield the monomeric radical parxylylene. On entering the deposition chamber, this coating is absorbed and polymerized on the substrate. Immersion comparison tests with acrylic, polyurethane, and silicone coats indicate superior results on the composites prepared with parylene coatings.[25]
- Gold-flashed palladium electroplated contacts have been evaluated with gold contact finishes on electronic connector hardware. Tests simulating a 20 to 40 year service life indicate that the palladium base electroplating is superior in maintaining stable contact resistance and wear resistance.[26] Hardness and thermal stability of the palladium are superior to gold, and the cost of palladium is about a third of gold.

VACUUM METALLIZING ON PLASTICS

Deposition of metallic vapors upon nonmetallic substrates, including plastics polymers, glass, and ceramics has become a major industry. Mirror-like decorative metallic films, formed by vacuum deposition, may be colored by the thin polymer film which is applied to seal and protect the metallic layer from abrasion. Microelectronic circuits, electromagnetic shielding, and integrated circuits are discussed in greater detail in later chapters. Packaging materials for food products and decoration, identification labels of metallized films which are applied by hot stamping, mirrors and toys, are representative consumer products which enjoy the advantages of composites of thin polymer films and thin metal films.

Aluminum metal is the material most frequently used with this technique and it is vacuum deposited in thin layers, usually from 1 μm down to 50 nm. The intact and coherent metallic film is fragile, requiring the toughness and abrasion resistance offered by the thin organic film overlay, as well as the support from the substrate.

Optical devices, in many instances, utilize metal vapor deposits (principally aluminum) on reflective surfaces. Formerly chemically deposited silver films were in vogue. Mirrors, eyeglasses, camera lens, and enclosed automobile headlight reflectors are popular consumer items which are made with metal/polymer composites. Large glass panels containing reflective metal vapor coatings are being used in new high rise building designs. The metal vapor films are sealed and protected by tough, weather-resistant polymer films. The composite of glass substrate, colored reflective metallic film, and polymer topcoat have won wide acceptance in architecture (Fig. 5-8) and underscore dramatically the success of metal/polymer composites.

The highly reflective glass used in building construction usually appears in the form of dual pane insulating glass. The two panes of glass are separated by hollow bronzed or aluminized spacers and hermetically sealed by polyisobutylene, silicone, or polysulfide formulations. Desiccants of silica gel and/or molecular sieves are inserted into the hollow metal spacers, and provide lifetime dehydration of the airspace (one quarter inch space or one half inch space are usually available). Vacuum metallized films, as described in this chapter, have been applied on the inside face of one of the glass panels. In this manner, the thin, 1 μm or less, reflective layer of aluminum, for example, has been sealed and abrasion protected with a clear or colored acrylic lacquer, and its presence in the sealed dual pane construction insures its stability (Fig. 5-8).

Liao has calculated the light transmission and attenuation for gold films deposited upon transparent substrates as a function of film thickness in angstroms (one angstrom equals 10^{-10} meter). A chart illustrating these results is shown in Figure 5-9. Thus for a gold film 80 angstroms thick, light transmission is about 80% at the peak of the visible light wavelengths.[38]

The processes of vacuum metallizing of plastics must anticipate the following factors:

A. Multiple parts produced by molding, casting, or fabrication are manufactured by batch processes.

- Surfaces must be thoroughly cleaned to remove mold, lubricants, and other contaminants.

Figure 5-8. Vacuum metallized glass panels are used on many new buildings.

- A lacquer coat is recommended to seal the pores of any surface, and provide a smooth, blemish free surface for vacuum metallization.
- Aluminum vapor deposited coatings can be made in many colors, which are applied outside the vacuum chamber. Batch processing vacuum equipment is shown in Figure 5-10.

126 METAL/POLYMER COMPOSITES

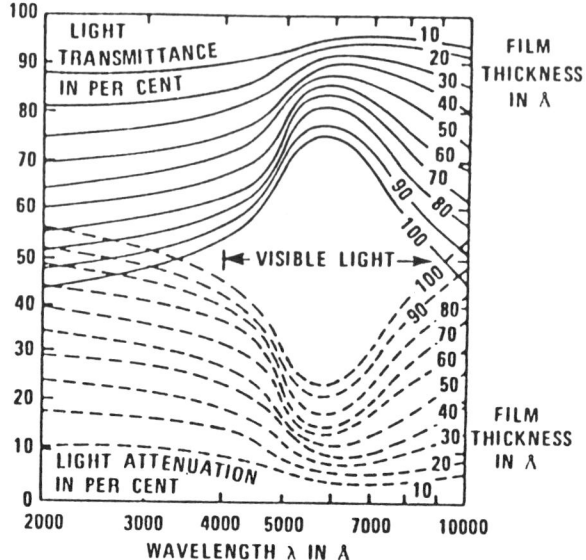

Figure 5-9. Light transmission through thin gold films. Data by S. Liao.[38] Courtesy of Prentice-Hall Publ. Co. NJ.

- Aluminum film must be sealed promptly with polymer coatings to minimize oxidation and to develop abrasion resistance.

B. Continuous polymer films of polyester, polyolefins, and polyimides are used for vacuum metallizing.

- The extrusion and high speed automation processes used for preparing polymer films provide a relatively contamination free film.
- Because of the need for metal vapor evaporation in high vacuums, and the large areas of coiled polymer films to be processed, large vacuum chambers are needed to accommodate the foils and the wind up and unwinding machinery. The entire operation, including materials supply and rewinding of the metallized foils, takes place under a high vacuum condition. The metal vapor must be concentrated along the width of the exposed foil surface. Before the roll is placed in the high vacuum metallizing chamber, a prior vacuum degassing is recommended.[27]
- Outside the vacuum chamber, the foils may be treated with colored coatings before rewinding. If the color of the metallized foil can be achieved in the molten metal alloy, the outside coloring step may be

Figure 5-10. Commercial vacuum metallizing chamber. Courtesy of High Vacuum Equipment Corp., Hingham, Mass.

unnecessary. Note: less than 20% aluminum in copper will give a yellow-gold tint to the foil.
- Metallized foils are sealed by top polymer coatings for abrasion resistance, and as noted above, colored impregnants may be used.

C. Large objects, which may require vapor deposited metal films, may function as their own vacuum chambers. The above outlined preparatory measures would have to be followed, and a sealed cover, to withstand the external stresses and to withstand a high vacuum, should be used. A good example of this unique situation is the aluminum vapor coating applied to the 200 inch glass reflective mirror of Mt. Palomar Astronomical Observatory in California.

Vapor depositions of metal coatings have been achieved by:
- Cathode sputtering using a high voltage discharge between electrodes (500 to 2000 volts) to propel the metal towards the object to be plated. The object functions as the electrode. A nonmetallic plastic body would require an electrically conductive surface.

- Volatilization of the metal by heat and vacuum, and deposition of the metal vapor upon cold surfaces. The quality of the metal coating is affected by an inadequate vacuum and contamination.
- Deposition has also been achieved by thermal decomposition or hydrogen reduction of metal compounds. This is a more costly procedure, but it may be required for special elements not accommodated by cathode sputtering or vacuum deposition.
- Chemical Vapor Deposition (CVD) methods have been used for deposition of high purity aluminum coatings on various substrates. Gaseous feed materials will yield solid deposits and gaseous by-products. High purity aluminum has been obtained by thermally decomposing the vapors of tri-isobutyl aluminum [$Al(C_4H_9)_3$] at 260°C. Plating rates of 8 μm per minute have been reported.[28]

To counteract galvanic action between metal objects being coated by vapor deposition of another surfacing metal, it is usually prudent to seal the metal object with a polymer coating before metal vaporization takes place.

Because of the popularity of the vacuum deposition of metals, most of the attention in the following paragraphs will be directed towards this technique. Narcus has described equipment necessary for the handling of multiple pieces, such as molded cabinets, jewelry, cosmetic packages, etc.[27]

The typical thickness of vapor deposited aluminum films range from 50 nm to 1.0 μm. The metal film mirrors the surface finish of the substrate upon which it is deposited, reflecting 80% of the incident light. Brilliant mirror-like coatings of metal can be applied to plastics and other nonconductors. The thin, uniform films that are formed are deposited inside a vacuum chamber.

Equipment Requirements. For effective vacuum metallization, particular concern must be shown for the ancillary equipment.

1. The design and selection of jigs and fixtures for holding parts are significant components of the equipment. Clamps must be positioned on the work so as not to interfere with the condensation and deposition of the metal. Special rotating fixtures for many small parts may be necessary (jewelry for example). The positioning and removal of parts is labor intensive and costly.
2. Metal evaporation (See chart of boiling points of various elements under vacuum, Table 5-3). Air molecules collide and interfere with metal vapor particles. Therefore high vacuum conditions are necessary. Helical coiled tungsten wire heating elements in contact with aluminum ribbing and wires, or alternatively small crucibles of tungsten, graphite, or molybdenum are used for supplying molten metal in the vacuum chamber (at

Table 5-3. Evaporable Metal Films*

ELEMENTS	LATENT HEAT OF FUSION (CAL./GRAM)	VAPOR PRESSURE (ATMOSPHERES)		
		10^{-3}	10^{-6}	10^{-9}
		BOILING POINTS (°K)		
Al	95	1782	1333	1063
Cu	49	1862	1391	1120
Au	15	2023	1510	1211
Sn	14.1	1857	1366	1080
Ni	71	2156	1646	1343
Pb	5.5	1230	889	698
Zn	27	752	559	449
Ag	26.5	1582	1179	952

*From *The Handbook of Chemistry and Physics* 67th ed. D-185, CRC Press. 760 mm Hg = 1 atmosphere
1 torr = 1mm Hg @ 0°C (STD. gravity)

10^{-6} atmosphere). Induction heating of the molten metal is generally practiced. As electrical conductivity is a function of the thickness of metal being deposited, continuous monitoring of thickness of foil coatings may be performed.

3. The key equipment to lower vacuum pressure to 10^{-6} atmospheres are the high vacuum pumps. For the early rapid removal of air, a large mechanical pumping unit achieves a partial vacuum. This is followed by diffusion and booster pumps which achieve a higher degree of vacuum, down to 10^{-6} atmospheres. High molecular weight liquids used in diffusion pumps replaced the moving parts of a mechanical pump. The liquid is brought to a boil and discharged through jet holes, collecting random air molecules. The vapors are condensed, the air discharged, and the diffusion liquids are recirculated. Low vapor fractions of hydrocarbons, silicone oils, or dibutyl phthalate are typical of the liquids. The performance of vacuum metallization is dependent upon developing and controlling the high vacuum required.

Decorative vacuum metallizing is generally less expensive than electroplating. This is one of the reasons why large quantities of polymer foils of polyethylene, polypropylene, and polyester are prepared for packaging purposes. The colored lacquers that seal and protect the vapor deposited metal films afford an unlimited choice of color selection for decorative purposes.

For imprinting identification and special markings on molded and extruded plastic articles (Fig. 5-11), foils of vacuum metal coated films are fed to hot stamping machines to coat protruding areas. Distinctive metallic symbols are impressed upon many consumer goods in this manner. Wasted material is found in those areas of the metal coated foil which do not participate in the stamping

130 METAL/POLYMER COMPOSITES

Figure 5-11. Signs imprinted with metal coated polymer foil.

and trimming operation. Heat activated adhesive film on the foil backing facilitates good adhesion to the substrate. For surfaces other than flat substrates, some foil transfer machines use a temperature resistant silicone pad to adjust to the moderate curves of molded plastic. These are the surfaces upon which the aluminum coated foils are to be impressed.

Several current patent and research references draw attention to unique metal/polymer composites, with metals applied by vacuum deposition processes.

1. High density polyethylene substrates were exposed to evaporated metals from electron beam bombardment. Qualitative examinations indicated that the adhesions from films of titanium, nickel, and chromium were high. On the other hand, adhesion was low for metallic films from aluminum, copper, silver, and gold. Photoelectron spectroscopy shows that high adhesion was aided by metal-polymer interaction, as well as by an increase in the strength of the substrate surface.[29]
2. The vacuum depositions of 82:18 (Co-Cr) alloy took place on a 15 μm polyamide film and the application of a magnetic alloy on a 7 μm poly(ethylene-terephalate) film are discussed in Reference 5-30. The metal application took place at an angle inclined to the moving film. This angle and the distance were critical. These developments were important to attaining curl free magnetic tapes.[30]
3. Vacuum deposition of metals in plasmas of organic monomers yield coatings for magnetic recordings. Ten μm thick polyester film was coated with up to 1000 angstroms of cobalt in this manner. The film, moving at a rate of 5 meters per minute, passes through a chamber with C_2F_4 at

.005 mm Hg pressure. Cobalt vapor from a 200 watt high frequency generator deposits a 200 angstrom coating of cobalt. The coated polyester film demonstrated improved flexibility and corrosion resistance over earlier types.[31]

4. Metal foil transfer films were prepared on a 12 μm polyester film coated on one side with a protective release layer. A 450 angstrom thick film of aluminum (vapor deposited) had holes perforated with multiple fine needles (holes too small to be visible). The aluminum layers were coated with an adhesive and transferred to an ABS molding. No flakes, separations, or cracks appeared after 48 hours at 90°C. Without perforations in the aluminum foil, flaking and cracks occurred.[32]

5. Thin films of metal deposited by vacuum evaporation were originally developed for optical elements in telescopes, binoculars, range finders, etc. Since the late 1940s, this technique has expanded into many other fields and vacuum deposited metal coatings on polymers supply composites for many industries. Many plastics are amenable to metallization by vacuum evaporation, particularly aluminum, copper, gold, and silver. Equipment for this process consists of a high vacuum pumping system which can achieve a vacuum of 10^{-4} to 10^{-5} mm of mercury, and a means for bringing the metal to its evaporation temperature (Table 5-3). In chapters 7 and 9, a number of applications of vacuum metallizing of electronic parts will be noted. Figure 5-12 illustrates schematically the vacuum coating of aluminum on polymer films.

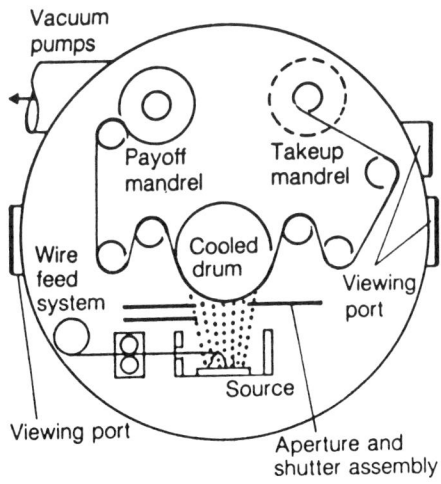

Figure 5-12. Continuous metallizing of polymer film. Adapted from Reference 13.

6. Thin metal films such as vapor deposited aluminum have been used to establish surface identifications on millions of plastic credit cards. Such films are difficult to duplicate or to alter. The application of thin metal films of the order of 1 μm or less to insulative surfaces are examined in a later chapter on electronic applications.

ABLATIVE AND INTUMESCENT COATINGS ON METALS

The protection of metal components from extreme high temperatures has been accomplished by developments in ablative coatings.[36] Two classes of ablative coatings may be noted.

Sacrificial Coatings. These were originally developed for the protection of nose cones and aerospace components during reentry into the earth's atmosphere from low orbital trajectories. The rapid deceleration from orbital velocities entails the expenditure of large amounts of energy which must be absorbed by ablative composite coatings. Considerable exfoliation takes place as gases expand the base material to provide thermal insulation. In this example, it is necessary to have a polymer with a high char strength. Phenol formaldehyde resins and furfuryl alcohol resins are among the commercial materials which have proven to be acceptable for rocket nozzle reentry temperatures.

High Temperature Stability Coatings. When high (and decomposing) temperatures are encountered, dimensional stability and mechanical integrity, as far as possible, is necessary. NASA specifications on ablative coatings for reentry spacecraft have concentrated on PTFE components of a shore diameter hardness of D55 to 70, elongation of 100 to 200% and tensile strength of 2500–3000 psi. Performance requirements dictate transparency to radio frequency transmission. Reentry temperatures have necessitated a minimum of alkali impurities which in the case of sodium and potassium is not to exceed 2.0 ppm. For maximum thermal stability, inorganics such as steatite and fabricated ceramic blocks (as used on the space shuttle) are superior.

Intumescent Paints. These have been used for many years and continue to provide thermal and fire resistance for structures. During conflagration, mass loss occurs at the front surface of a body, and is accompanied by outgassing. There are discontinuities of density along with the temperature gradients. A recent report establishes models for analysis.[33]

Intumescent polymer coatings for metals have been used for specialized purposes to protect metal bases from intense heat. Specific film forming polymers are formulated to react under extreme heat to form a foamed product that shields and protects the metal product which has been covered with the coating. In practice, the intumescent coating contains chemicals which would cause the coating to exfoliate, foam, and exhibit char resistant characteristics when exposed to the high temperatures of an intense fire. Many polymers including urea formaldehyde, polyurethanes, polystyrene, polyvinyl chloride, and others have been used in intumescent coating protection. They generally experience a 2 to 5-fold expansion to establish an insulating shield and remain as a rigid thermal insulation. Because of the prevalence of halogenated derivatives in their formulation, the fumes are quite toxic.[34]

REFERENCES

1. Nicholson, J., *New Scientist*, Vol. 110, p. 42, (May 29, 1986).
2. C.A. 106,175,540f, H. G. Reville, A. Pardies Soc. Ind. Aerospace (France), (1986).
3. Delmonte, J., *Concrete Jl.*, Vol. 47, p. 12, (June 1949).
4. *Physical Properties of Formulated Polyvinyl Resins*, B. F. Goodrich Co., (1988).
5. Vassiliou, E., E. I. duPont Co., U.S. 4,645,692, (February 24, 1987).
6. C.A. 106,103,931r, Daido Steel Sheet Co., J.P. 61-236-869, (October 22, 1986).
7. C.A. 107(16)-135,542b, S. Onishi J.P. 62-109-827, (May 21, 1987).
8. Takakura, M., et al., (Nissan Chemical) U.S. 4,604,303, (August 5, 1986).
9. Scala, L. and Phillips, D., (Westinghouse), U.S. 4,003,812, (January 18, 1977).
10. Rosen, T. and Lane, T., U.S. 4,554,185, (November 19, 1985).
11. Anon., "Powdered Coatings," *Materials Engineering*, 103:42, (October 1987).
12. C.A. 84-75802F, Kansai Paint Co., 1976, J.P. 75,142,641, (November 17, 1975).
13. Himler, D. and Alzapiedi, J., "Preparing Plastics for Plating," *S.P.E.*, Vol. 29, p. 48, (June 1973); and *Modern Plastics Encyclopedia*, p. 349, (1988).
14. ASTM-B-727-83, *Preparation of Plastics for Electroplating*, p. 633, 365, (1985).
15. C.A. 106-106554F, IBM, European Pat. Applic. 206,145, (December 1986).
16. C.A. 106-157214e, Toray Ind., (1987).
17. Waggoner, J., "Electroplating and Sputtering," *Mod. Plastics Encyclopedia*, p. 349, (1988).
18. Hartsing, T. et al., U.S. 4,588,623 to Union Carbide, (May 13, 1980).
19. Scheer, G. R. and Preston, J., (Battelle Dev. Corp.) U.S. 4,597,836, (July 1, 1986).
20. C.A. 105(18) 154,219b, Toppan Printing Co., J.P. 61,150,299, (December 24, 1984).
21. C.A. 106-158-020Y, Nippon Denki K.K., J.P. 61,177,398, (August 9, 1986).
22. USSR, C.A. 106(18), Section 42, (1987).
23. C.A. 106-105426d, E.I. duPont de Nemours, European Patent Appli. 191,248, (August 1986).
24. C.A. 104(20), 176,660V, Y. Tanoka, Mitsui Mining Co., J.P. 60,255,995, (1986).
25. Humphrey, B., *Parylene Coating Bulletin*, NOVA Tran Corp., (1988).
26. Micklus, S., (AT&T Nassau Metals), "Palladium Plating," p. 41, (May 1986).
27. Narcus, H., *Metallizing of Plastics*, New York: Reinhold Publishing Co., (1960).
28. Aerospace Materials Report, ML-TDR-65-114, (April 1965).
29. C.A. 105(18) 154,237F, Bodae, P. (Sweden), "Adhesion of Evaporated Metal Films to Polyethylene," (1986).
30. C.A. 105(18) 154,278V, Matsushita Elec. Ind. Co., J.P. 61,99,935 and 938, (May 18, 1986).
31. C.A. 106-86290u, Hitachi, J.P. 61,214,138, (September 24, 1986).

32. C.A. 105(18) 154,849g, Reiko Co., J.P. 61-72,582, (April 14, 1986).
33. Buckmaster, J., *Intn. Jl. of Engrg. Science,* 24:263, (1986).
34. Tashleck, I. and Valenziano, P., U.S. 4,588,523, (May 13, 1986).
35. Inui, T., et al., Toyo Kohan Co., U.S. 4,614,691, (September 30, 1980).
36. Sixth Sagamore Materials Research Conference at Syracuse University—Ablation Protection, p. 431, MET-661-601, (August 1959).
37. Herman, H., "Plasma Sprayed Coatings," *Scientific American,* 259:112, (September 1988).
38. Liao, S. *Microwave Devices,* Englewood Cliffs, NJ: Prentice Hall. (1980).

6

METAL/POLYMER STRUCTURAL COMPOSITES

Combinations of organic polymers and metals have fulfilled important roles as structural components, and have been used for very visible applications in the automotive and aerospace industries. In these industries metal/polymer composites have met physical design requirements with significant savings in weight over corresponding all metal components. To aircraft designers these savings are equated with higher speeds and greater payloads. For automotive designers there are manufacturing advantages to be gained in the use of polymers for joining, in one assembly, multiple metal components.

High-strength fibers in organic polymer matrices have been used extensively since glass fibers were introduced in the 1930s. Diverse glass fiber reinforced components are accepted by industry. Polyaramid fibers and polyethylene fibers have, in some instances, supplanted glass fiber reinforcements because of the demand for high-impact resistance. Structural requirements which may have demanded that the high modulus of elasticity of steel be used in composites, have found carbon-graphite unidirectional fibers and silicon carbide reinforced aluminum wire to yield acceptable polymer composites. Other ceramic fibers of aluminum oxide and nonmetallic boron deposited on a tungsten filament also offer structural advantages.

Various fiber combinations of organic, ceramic, and metallic filaments are available to form hybrid mixtures. The roles that these mixtures are to assume are still being defined, and the creation of these mixtures has led to great interest in new applications for structural composites of polymer matrices reinforced by ceramic fibers, polyaramid fibers, carbon fibers, silicon carbide fibers, metal-coated organic fibers, gold wires, reinforced aluminum wires, and special hybrids.

Interest in structural composites is strong because of new developments in the following areas:

1. The controlled carbonization of organic fibers which yield high heat-resistant fibers.
2. New organic polymers which provide improved stiffness and heat resistance.

3. New light-weight metallic alloys of aluminum and of magnesium which exhibit improved physical properties. These alloys function as matrices in structural composities.
4. Ceramic fibers which are capable of functioning as reinforcements to aluminum and magnesium, and which offer competition to silicon carbides for such uses.
5. Hybrids of metal fibers, organic fibers, and ceramic fibers.
6. Development in light-weight structures involving honeycomb cores of metal and plastics, and facesheets of fiber-reinforced polymers or metals.

In this chapter, there are several techniques which will be examined for establishing the basis for metal/polymer structural composites. They will include:

1. Reinforced polymer coatings on metal applied by polymer impregnated tapes and fabrics.
2. Reinforced polymer coatings on metal applied by filament winding and pultrusion techniques, and resin treated fibers.
3. Impregnation of porous metal compacts, wire wound electric motors, wire ropes, and cables with solvent-free liquid polymers.
4. Lamination of metal sheets with polymer adhesives, films, or foams for sound absorption, reduction of vibration, and thermal insulation.
5. The roles of metal matrices in polymer/metal composites, including graphite fibers from organic polymers and ceramic fibers.

Considerations of metal/polymer structural composites are influenced not only by the properties of the finished part, but also by manufacturing advantages. For example, the composite framework for an automotive chassis was recently fabricated at Budd Company by resin transfer molding. Six molded parts, including metal inserts, eliminated over one hundred metal components that had been required in an earlier design.[1] The prospects for lighter weight and noise abatement are obvious. Interests in these metal/polymer composites are further stimulated by factors such as:

• Tooling costs for reinforced plastic/metal composites (N.B., the tools, jigs and fixtures may be also constructed in part of reinforced plastics) will be considerably less than metal tooling costs. Production time for preparing nonmetallic tools is much less than for machining metal tools, when accurate models are available.
• Resin impregnated fabrics are available from well established preppegers who have specialized in the selection and controlled impregnation of fibrous materials suitable for reinforcing purposes. Admittedly, depending

upon production volume, a precision machined steel tool will outperform a nonmetallic tool in large production, but the availability of the nonmetallic tool will be days or weeks ahead of machined metallic tooling.

Nonmetallic tooling has been well documented in other publications.[2,3,4] It has enjoyed extensive use as a technique for the aerospace industry which has a more limited production, as compared to the automotive industry. Metal inserts in plastics tooling are common, and represent successful composite tooling. A typical example is the forming die for a sheet metal fan illustrated in Figure 6-1. Metal blocks and the tool framing have been properly aligned. The critical shape and curvature of the contoured blades have been reproduced in a tough castable, filled polymer, from a precision model. The plastic surfacing medium on the metallic blocks is contoured to the required fan blade profile. This is an example of metal/polymer composite tooling, and its good structural behavior.

Metal/polymer tooling takes advantage of the innate qualities of the metal chase and frame for high strength and the resiliency of a polymer layer to accommodate sheet metal variations. Should profile changes be required, the nonmetallic layer can be replaced on the metal base.

Jigs and fixtures for drilling, welding, assembly, etc., are prepared from

Figure 6-1. Polymer faced metal tooling for forming a fan. Courtesy of Bell Aircraft Corp.

138 METAL/POLYMER COMPOSITES

fiber reinforced plastics, and the wear and stress positions are supported by metal inserts. These tools are examples of metal/polymer reinforced composites. Materials of composite tooling are diverse and include: epoxy glass laminates, polyurethane and silicone flexible pads, steel and aluminum filled casting materials, graphite blocks, and metal inserts for heat, impact, and wear resistance.

Structural applications of metal/polymer composites are found on aircraft and automotive vehicles (Fig. 6–2). The list of uses is very large and there are indications that the use of plastics materials for weight savings will increase. It has, therefore, been recognized that certain performance functions are handled best by polymers, and that other physical and high temperature functions are best handled by metals.

There have been many successful structures of polymer composites using fiber reinforced polymers such as epoxy resins, polyetheretherketone (PEEK), polyimides, bismaleimides, and others—which in combination with honeycomb construction and/or fibers of graphite, glass, or polyaramid—have achieved

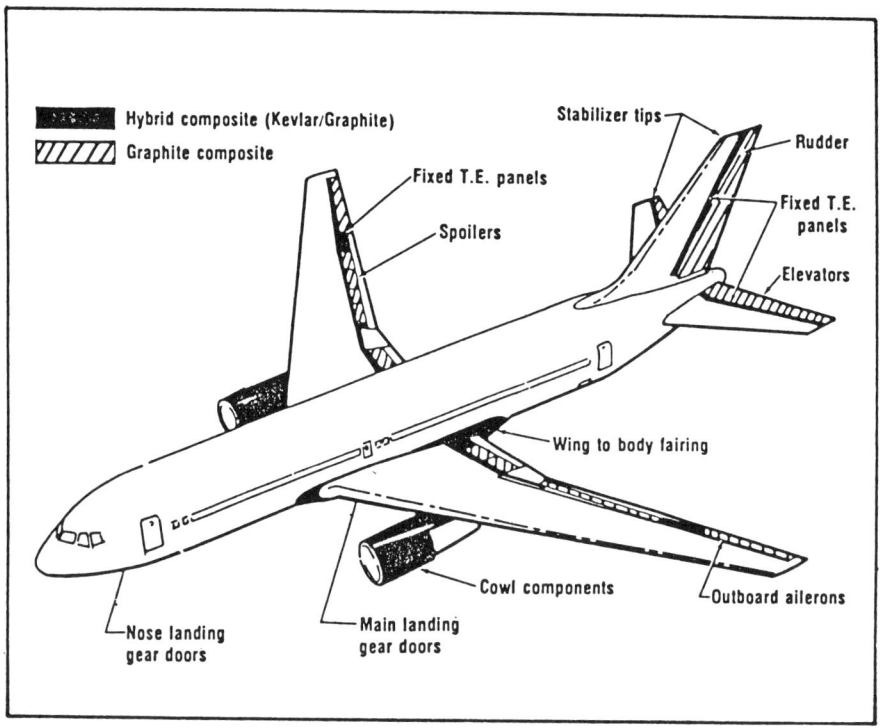

Figure 6–2. Typical structural composites on a Boeing 767.

substantial weight savings in aerospace vehicles. Metal matrices used with the reinforcement of metallic filaments, silicone carbide filaments, aluminum oxide, and metal-coated graphite, exhibit excellent performance in structures. These represent advanced metallic structures for high temperatures. The synergism of metal/polymer composites is emphasized in the examples which follow.

For specific details on a large number of resin matrices, metal matrices, fiber reinforcements, ceramic reinforcements, and environmental influences on structural composites, a recent publication of the American Society of Metals (ASM) is suggested.[57] It is imperative that more data on the physical properties of metal/polymer composites be obtained. This data should be obtained when composites are under stress, and concomitantly exposed to real-life adverse environments.

There are seven categories in which the examples of metal/polymer structures may be classified (the examples appear immediately after the categories):

1. Weight Savings: The lower specific gravity of reinforced polymers may allow substantial weight savings when they are designed in association with metals.
2. Vibration and Noise Abatement: Metal panels surfaced with more resilient polymer are more acceptable because of their ability to curtail noise and absorb vibration.
3. Improved Impact Resistance: From armor plate to metal sheets with resilient cores, improved impact resistance is observed.
4. Improved Formability: Metal-polymer-metal laminates have been developed with the objectives of improved processing and forming of metal sheets.
5. Improved Appearance: Metal sheets may be functionally enhanced with polymer surfaces which offer improved appearance and superior corrosion resistance.
6. Miscellaneous Improved Performance of Metal Polymer Reinforced Composites: Polymer tapes will improve the durability of sheet metals. Sprayed metal alloys will enhance the abrasive resistance of polymer tubes and rolls.
7. Structural Base for Electronics: The manufacture of laminated thermosetting circuit boards permits the assembly and orientation of many electronic components.

1. Weight Savings. Helicopter blades necessitate important design and weight distribution considerations. Maximum stresses appear at the hub of the rotating blades and at the areas of attachment to the engine. Minimum weight needs arise at the tips of the blades. While a solid *chunk* of metal may be machined

140 METAL/POLYMER COMPOSITES

to a required weight distribution, costs of manufacturing can be reduced, and enhanced performance attained through adhesively bonded laminates of high strength aluminum alloy and laminated graphite, glass, or polyaramid fibers, specifically aligned to take advantage of weight and strength considerations.

Supersonic aircraft, including the shuttle aircraft which is placed into earth orbit, and transport airplanes for high speed intercontinental travel utilize many classes of composites. High strength and light weight considerations lead to the use of metal/polymer composites. The first space shuttle utilized composites on the aircraft exterior and interior. Under the compelling requirements of high strength, light weight, and environmental design efficiency, the roles of composites will multiply.

Metal-thermoplastic-metal laminated composites were developed to provide light weight laminates having improved service properties for construction applications. These laminates were used for exterior and interior building panels and transportation vehicle body panels. Thermoplastic core layers were selected from partly crystalline polyamides and polyesters. The composites attained heat distortions above 130°C, a maximum weight of 9.76 kg/m^2; and a coefficient of linear expansion less than 10^{-6}/°C. The minimum core thickness was .483 mm.[12] Earlier patents by Bugel and Zunich suggested a similar approach.[13]

Light weight composite sheets for automobile bodies are prepared from foamed polymer sheets, laminated with polyester glass fabrics, and 0.2 mm thick steel plates using a polyester adhesive. The panels were pressed for 6 minutes at 85°C.[15]

Honeycomb reinforced panels, with aluminum reinforcement, which are resistant to hostile fluid environments have been created. To achieve this product the aluminum honeycomb core foil is anodized in phosphoric acid and coated with a corrosion resistant primer before the core is made.[27] Figure 6-3 shows the elements of the polymer/metal composite in the honeycomb assembly.

Light weight polymer/metal composite sheets have been developed from polyurethane foam, plus a 20 mm glass wool (bulk density .02 g/cm^3), faced with steel plates 0.2 mm thick and coated with a polyester adhesive. The sheets were pressed for 6 minutes at 85°C in the manufacture of automobile panels.[28]

2. Improved Vibration and Noise Abatement. Door Assembly of Polyurethane Molded Over Metal. The outer skin of a new door design is reaction-injection molded polyurethane (a low viscosity two component system) applied to a load bearing metal skeleton. Glass fiber reinforcements are also used. This integrated composite molding is dimensionally stable and provides an anchor for the lock, hinges and door handle. A finish coat of a flexible polyurethane is applied to the door assembly.[7]

METAL/POLYMER STRUCTURAL COMPOSITES   141

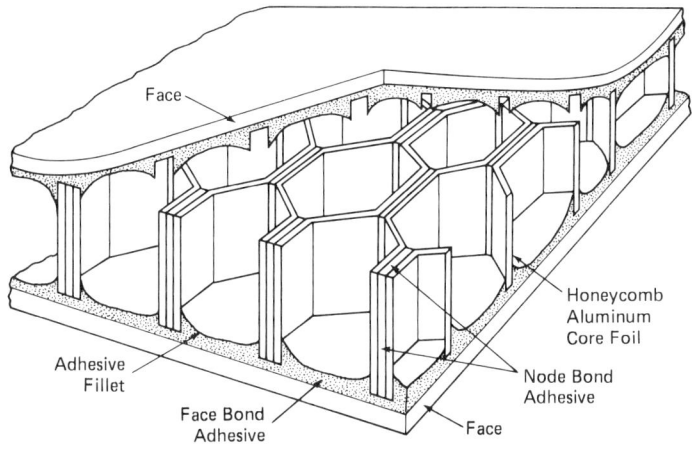

Figure 6-3. Aluminum foil used to improve the moisture resistance of a honeycomb assembly. Courtesy of American Cyanamid Corp.

Alcoa's Aerospace Sheet and Plate Division has formed a group to market *Arall* laminates, a high strength metal/polymer composite that is formed of aluminum-polyaramid laminates. It behaves like aluminum sheet and is recommended for aircraft applications where good fatigue resistance and acoustical dampening are required. In the event of fatigue cracks in the aluminum, the aramid fibers tend to limit the spread of such cracks and reduce stress intensity (Fig. 6-4).

Using uniaxial prepregs, and a 3/2 lay-up (1.3 mm total thickness), Alcoa reports a 35% increase in tensile strength in the longitudinal direction. Data are reported for Al alloy 7075T6 and Al alloy 2024-T3.[10]

Injection molding rubber compounds into spaced metal sheets is a technique used for forming laminated composites of metal and rubber.[17]

A thermoplastic composition was prepared from coal tar pitch, a terpolymer of ethylene, vinyl acetate, and ethylene acrylic acid copolymer, with a bottom metallic layer (steel, galvanized steel, or aluminum) at least 0.25 mm thick. Many variations of the thermoplastic composition are described in the patent

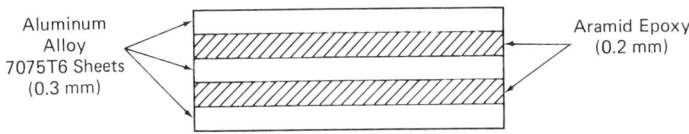

Figure 6-4. Cross-section of Alcoa's five-ply laminated composite.

Table 6-1 Aluminum-Polyaramid Epoxy-Laminates[10]

AVERAGE MECHANICAL PROPERTIES (LENGTHWISE)	7075/ARAMID COMPOSITE	2024/ARAMID COMPOSITE
Ultimate tensile *KSI*—(MPa)	112 (784)	102 (714)
Tensile yield strength *KSI*—(MPa)	95 (665)	51 (357)
Elongation (%)	0.6	1.3
Tensile modulus of elasticity *MSI* (GPa)	10.1 (70.7)	9.5 (66.5)
Density (g/cc)	2.3	2.3

which discusses its application, because of its good weather resistance, to industrial roofing and sidings. Thin layers laminated between steel sheets possess good vibration dampening characteristics. Preferred coal tar pitch had a ring and ball softening point between 100°C and 150°C. In preparing the laminate, the steel sheet is heated to about 210°C and is passed with the adhesive mastic layer for bonding through a nip-roll.[23]

Composite sheets of polymers and metals are not always fabricated in a press under heat and pressure. Sound and vibration damping materials are applied to interior panels of commercial airliners. Polyurethane foam and a special aluminum foil backing are applied with a resilient adhesive to metal skins, where they sharply reduce interior noise and vibration-caused failures on the skin. The use of polymer impregnated glass fabrics or high strength polyethylene fiber *(Spectra)* to counteract cracks in metal skins was discussed in the previous chapter, in the section on reinforced coatings. In view of recent fatigue failures on the skins of old aircraft, this reinforcement is an important safety measure. Layered composites described in this paragraph are also reported as being used as dampening panels in building construction to introduce a measure of vibration control.

3. Improved Impact Resistance. Armor Plate. Solid sheets of high strength steel (armor plate) have been considered the most effective materials for absorbing the energies of impacting bullets and projectiles. Recent designs in advanced composites have utilized shock absorbing polyaramids, superstrength polyethylene fibers *(Spectra)*, and glass fabric laminates to distribute the impacting energies over a greater area. The fracture toughness of polyaramid and carbon fiber epoxy composites have been significantly increased by a thin coating of polyurethane applied to the fibers.[53] Similar results were not obtained with silicone coatings.

The success of *body armor* (designed to stop bullets) manufactured with polyaramid fibers and high strength polyethylene fibers, has been noteworthy.

For stationary shields, the intercalation of steel plates and semiresilient adhesive layers behaves much like the safety glass sheets laminated on either side of semiflexible polyvinyl butyral or polyurethane resins. Opportunities for material designs of metal/polymer composites in armor plate exist in many guises.

High heat structural metal polymer laminates are obtained by the lamination of steel or aluminum plates with a polymeric resinous core consisting of the blend of an engineering thermoplastic resin with certain block copolymers. The block copolymers[8] are preferably of the species of polystyrene-polybutadiene-polystyrene (SBS) and polystyrene-polyisoprene-polystyrene (SIS). The block copolymers are blended with engineering thermoplastics possessing strength, good modulus, high impact resistance, and environmental stability. Glass transition temperatures fall in the range of 150°C to 300°C. Polybutylene terephthalate is a typical satisfactory thermoplastic. The polymeric resinous material tightly adheres to steel skins about 0.1 to 1 mm thick with excellent adhesion and high heat capabilities.[9]

A polyolefin compound has been developed by a steel corporation for use in laminate steel-plastic core composites. Properties developed in the core required the use of 70% by weight of an isotatic crystalline polypropylene, and for improved impact 30% of propylene-ethylene-diene terpolymer. The formulation was filled with an inorganic filler selected from calcium carbonate, mica, and silica. The composite sheet with steel face plies was designed for high temperature baking of automotive finish.[11]

4. Improved Formability. Precoated iron sheets which could be bent, drawn, or stamped were prepared from polyester powder coatings containing .05 to 3% benzoin. The coatings were cured rapidly close to a gas heated plate (600°C surface temperature) for a few seconds.

A metal-plastic-metal structural laminate has been developed for formation into useful articles having compound curves. The polymer core was selected from homo-polymers and copolymers of ethylene and polypropylene, preferably irradiated for improved creep resistant properties and thermal stability. The core thickness was specified from 10 to 45 mils, and metal skin layers 5 to 15 mils, with total laminate thickness 30 to 60 mils. Aluminum alloys were mentioned in the patent claims, as was the use of intermediate acrylic acid adhesives.[21]

Alkyl acrylic polymers are grafted to polyvinyl chloride to form a polymer sheet 50 μm in thickness. This polymer was roller laminated to a zinc coating (0.35 mm) and the combination fusion bonded to a steel plate. The specific acrylic modification prevented stress whitening during processing. The laminate offered a corrosion and weather resistant product.[24]

Steel and aluminum composite sheets were prepared by laminating to their surfaces a biaxially oriented polyester film of polyethylene terephthalate. The metal sheets were covered with a double layer of hydrated chromium oxide (8 to 20 mg per m^2) and a lower layer of chromium (3 to 50 mg/mm^2). An epoxy resin adhesive with a polyurethane resin cure agent was used. Sheets were preheated to high temperatures to facilitate rapid cure.[22]

5. Improved Appearance. The preplating and/or precoating of sheet steel at the mill are measures for effecting cost savings for customers, who otherwise would find it necessary to set up their own coating facilities at greater expense. Specialty polymer coatings offer embossed designs, colored finishes, and special surface textures. New trends in the use of powdered metals at the steel mill presage increased activity in preparing metal/polymer composites. Polymer coating of steel coils at the mill is on the increase.

The electroplating and polymer foil coating of steel plates were covered in Chapter 5. Zinc-coated steel has been prepared for use on American made cars. The coated steel could be the base upon which polymer coatings are applied. Advances in the quality of zinc coatings and their paintability have been made with 5% Al, mischmetal (.04% La—.04% Ce) zinc coatings.[6] This zinc alloy has also been used for the hot dipping of the steel.

Manufacturers of table tops have used decorative inlays of metal stampings on the surfaces of laminated thermosetting plastic sheets. Light colored urea formaldehyde and melamine-formaldehyde resin systems, usually impregnated into alpha cellulose sheets have been used as the base upon which the metal stampings were placed. There is always the chance for movement of the inlays under the heat and pressure of the lamination process, mechanical positioning devices or locating pins are necessary. The use of additional technology such as silane coupling agents for enhancing resin to metal bond, and metallic electrocoating on the metal inlays would improve the quality of the composite.

Steel plates are coated with aluminum powder and heated to fuse the metal powder into a 25 μm coating. This is followed with a coating of an acrylic resin powder which is baked 20 minutes at 180°C forming a clear coating with good gloss.[25]

6. Improved Performance of Composites. Reinforced metal adhesives. Solvent free polymer adhesives, such as liquid epoxy systems, when applied to a thin glass fiber veil, and then applied and cured as an adhesive layer, develop considerably higher lap shear strengths between metal surfaces than the unreinforced adhesive. The reinforced adhesive assemblies exhibit an improvement in lap shear strength of up to 50%. The polymer fabric reinforced adhesive/metal combination forms a structural composite of considerable merit.

Naval ship designers have been aware of the advantages of the new external rib profiles on the hulls of racing ships. The success of United States yacht designers in capturing first place in the world competition held in Australia

(1987) is due to the skills and experiences of the crew and the inclusion of novel external rib profiles.[5] The manufacture of the rib profile is a role for extruded and shaped plastics. When these plastics are bonded or laminated to metal hulls they contribute to higher speeds. The riblets, placed parallel to the flow, are triangular in shape and reduce turbulence in the layer of water 0.25 mm from the surface. Preventing eddies at that distance from the surface decreases the turbulence 8 to 10 cm from the surface where it is most destructive.

Carbon fiber reinforced phenolic resin laminates are wrapped around cylindrical mandrels and cured with heat. They are plasma sprayed with Al-Cr-Mo alloy to a thickness of 0.1 mm. Abrasion resistant composite rolls are manufactured using these laminates and used in paper making machinery.[14]

Bonding of steel to polyethylene sheets was aided by the use of a polymer resin mixture of ethylene-vinyl acetate copolymer, polyamide, ABS polymer, and slaked lime.[18] Other references make note of corona discharges to improve the bonding qualities to the polyethylene surface.[19]

Bitumen (62 pbw) is mixed with calcium hydroxide (36 pbw) and blended with (2 pbw) ethylene-vinyl acetate copolymers. The product is cooled, milled to a finely divided form and applied to the surfaces of reinforced steel for use in concrete. The formulated coating powder was in the range of 100 to 300 μm. Coatings on reinforcing steel are particularly valuable in marine environments, where the corrosion of steel results in swelling of the steel surface, placing stress on the concrete in which it has been embedded.[20]

7. Structural Base For Electronic Circuit Boards. High performance large laminated sheets of copper for manufacture of circuit boards. Full size sheets measure 120 cm × 90 cm and thicknesses of .075 cm to .30 cm. For example, copper clad sheets are prepared to meet MIL-P-13949 specifications. Salient characteristics of these metal polymer composites appear in Table 6–2.

A flexible printed circuit board was prepared from a blend of ethylene-methacrylic acid copolymer and saponified ethyl vinyl alcohol at 250°C. This film was hot pressed between a copper foil and a heat resistant polyether ketone film to form a flexible printed circuit board.[26]

Polymer/metal composites are used in conduits and pipes for transport of liquids and gases. Corollary to the subject of composite pipes are the metal

Table 6–2 Properties of Copper-Epoxy-Glass Laminates[16]

Peel strength	.72 kg per cm at 290°C
Volume resistivity	3×10^9 megohms-cm
Surface resistivity	1×10^8 megohms per square
Arc resistance	120 secs
Glass transition temperature	170°C

fittings and couplings associated with their installation. The thermoplastic pipe material will permit adhesive assemblies in most instances. For more physically demanding installations, fiber reinforced tapes of glass or polyaramid may be adhesively bonded and wrapped about pipe junctions and unions. Not only will the joints be mechanically reinforced, they will also be better sealed and capable of offering resistance to root penetration if located in the ground.

The aerospace industries and communication industries have spawned many new technologies for developments of polymer/metal composites. Two of the technologies which are discussed in this chapter are filament wound structures and pultrusion, both of which involve novel manufacturing techniques. The versatility of the methods allows for varied material selection and distribution. Patents for the use of silicon carbide whisker additions to aluminum have resulted in remarkably high strength aluminum alloy wires which are necessary for some composite applications. Polyacrylonitrile (PAN) fibers, when carbonized and then oxidized under close temperature control, are the basis of much of the carbon/graphite fibers used for reinforcement of structural composites.[29] PAN fibers, referred to in patent literature as carbon or graphite fibers, have established their importance in polymer composites.

FILAMENT WOUND STRUCTURES

Resin bonded filament wound structures produced from glass filaments have enjoyed success particularly in the fabrication of internal pressure vessels. Polyaramid filaments, with a specific gravity lower than glass, are important when weight savings are a prime concern. These structures belong to a class of reinforced plastics which have developed a self-sustaining market. The techniques of filament winding of fibers, can be designed to prepare tubes, columns, beams and enclosures. There are specific situations in which containment of metals within a filament wound resin-bonded structure will be of advantage.

The development of cryogenic fluid containment tanks for space flight vehicles, started with glass fibers and continued later with fibers of boron, graphite, and polyaramid (poly-phenylene terephthalamaide—duPont's *Kevlar*). Figure 6-5 depicts a space shuttle vacuum container, which uses filament wound components.[30] The space shuttle cryogenic tanks of oxygen and hydrogen are used to generate the orbiter's electrical power and furnish oxygen for the crew compartment. Each tank consists of two concentric metal spheres for containment of the fluid. The oxygen tank having an outer diameter of 99 cm and the hydrogen tank having an outer diameter of 119 cm. The total static supported weight is about 408 kg for the oxygen tank and 82 kg for the hydrogen tank. Twelve filament wound glass/epoxy composite straps are used to support the pressure vessel from the outer shell. The nonmetallic composite strap supports minimize the conduction of heat from the outside to the inner-

OXYGEN TANK

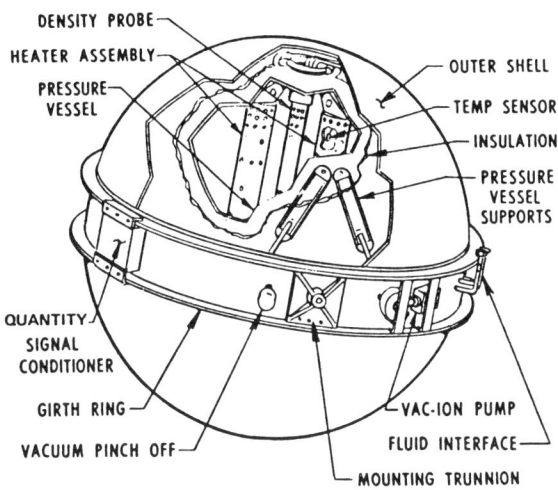

Figure 6-5. A space shuttle power reactant storage assembly of oxygen tank. Courtesy of ASTM and E. Morris, *Structural Composites,* Azusa, CA.[30]

shell of stored cryogen. They are located within the vacuum space between the spheres.

Among the design principles which are accommodated by filament winding are:

1. Thin shells of costly metals, such as stainless steel, molybdenum or tantalum, may function as the mandrel about which the reinforcing fibers are wound.
2. Metal fittings of steel, copper, or aluminum alloys, essential for threaded assemblies, may be located within the fiber wound structure, while it is being wound, without weakening the fiber structure by cut-out areas for their insertion.
3. Shielding of electromagnetic radiation, as discussed in Chapter 7, can be accomplished by metal shells or metal wires which replace part of the glass or polymer fibers during the winding process.
4. Filament winding is not necessarily limited to concentric shapes, and equipment may be designed for multi-axis rotation during the winding process.

Where glass fibers are used during filament winding, typical weights would be approximately 80% glass and 20% bonding resin (epoxy or polyester) in the final structure. The catalyzed resin is applied to fibers from a liquid bath. Tension applied to the fibers forces out excess resin and air, allowing each successive layer to rest on solid material. Helical winding of adjacent layers achieves the needed shear strength in the structure. Infrared lamps supply heat for curing resin during the winding process, or at the completion of the fabrication process. Filament winding techniques are applicable for sophisticated metal/polymer composites.

The physical properties of fiber-reinforced resin-wound structures are superior to those prepared by platen pressing or molding. The modulus of elasticity in tension or flexure is usually double the modulus of flat panels of reinforced resin composites, on which no tension on the filaments is maintained during the cure, and where the volumetric loading of the fibers is less.

PULTRUSION

The pultrusion process is ideally suited to the manufacture of continuous lengths of polymer, metals, and ceramics. The equipment is designed for physically engaging and pulling assemblies of fibers, filaments, tapes, and mats through a liquid resin impregnating bath, and into wiper profiles to remove excess resin and air. This is followed by curing to set the resin polymer in heated dies, or the material is cooled on a continuous basis if it is a thermoplastic. High strength metal wires, including silicon carbide fiber reinforced aluminum, boron, graphite fibers, quartz fibers, aluminum oxide, glass fibers, polyaramid fibers, and silicon carbide fibers, as well as extruded metal profiles are some of the materials which can be used in the pultrusion process. For hollow shapes, a mandrel is inserted ahead of the resin tank; it extends through the die for curing the composite. If the design includes the mandrel, so much the better. Because of the tension and orientation of the continuous fibers by the pultrusion machine, the magnitudes of the physical properties of the pultruded composite rods are higher in the parallel direction than in the direction perpendicular to the main axis of pultrusion. Figure 6-6 illustrates an exploded view of different materials being compressed during pultrusion. This is similar to procedures used in the manufacture of transoceanic cables for communications.

The lower strengths in the perpendicular direction may be compensated for by the inclusion of braided tapes applied over the core structure, or by additional rotational twisting measures applied to the whole assembly. The creels for supplying the filaments (roving) are located at convenient intervals to lead rovings to the resin mix tank. Care must be exercised to prevent the multiple rovings from scraping across one another. For nonconductive filaments, electrostatic charges would occur and create processing difficulties.

METAL/POLYMER STRUCTURAL COMPOSITES 149

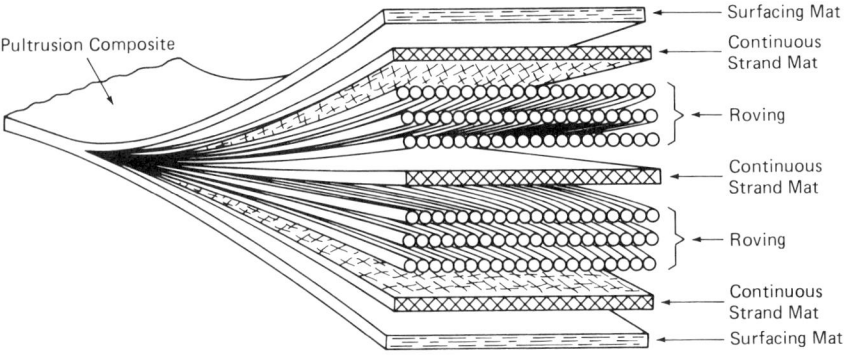

Figure 6-6. Diverse materials consolidated by pultrusion. Courtesy of L. Meyer.[55]

A group of ten 275 kv insulators with pultruded glass reinforced cores which had been in service were examined in various states of electrical and mechanical damage. Progressive electric damage started at the metal end fittings and proceeded along the interface of the core and the shedding. Tension sheds with high mechanical loads suffered cracking which appeared to have admitted moisture.[31] Also of interest is the fact that deformation of insulators under mechanical loading increased in the presence of an electrical field.

Not all polymer reinforcements for metal cores and special profiled inserts need be applied by the pultrusion process as there are other coating alternatives. The extrusion of insulative plastic sheaths (such as polyvinyl and silicone resins) over another base, such as metal or metal wires is a common practice. A tension force is applied to the entire composite assembly. Another alternative is the application of a braided tubular fiber stockinette textile over the metal core. The fiber reinforcement, usually of glass, aramid, nylon, high strength polyethylene, or polypropylene, can be woven about the metal core. If the tubular braid has been prefabricated, it may be manually pulled over the core. The innumerable flexible garden hoses used by householders are extruded of synthetic rubbers and rubber-like thermoplastic polymers. Some of the more expensive and durable hoses have a protective internal wire sheath of metal wires, which is also encased in an exterior plastic cladding (Fig. 6-7).

Kevlar polyaramid fibers are used successfully in reinforcing hoses (which maybe made by pultrusion processes using conveyor belts). The finished products are much lighter in weight than those produced with steel fiber reinforcement. *Kevlar* fibers also perform well in automobile tires (Armstrong brand).

Pultrusion and extrusion processes are able to develop multilayer polymer/metal composites to meet stringent physical requirements. For example, fire hoses which must withstand the weight of automobiles, which may run over

150 METAL/POLYMER COMPOSITES

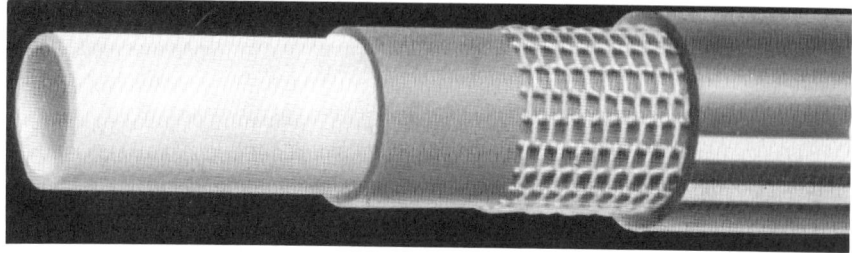

Figure 6-7. Metal braid reinforcement used in high-quality extruded and pultruded garden hose.

them, are prepared this way. Experiences learned in these applications may suggest creative designs requiring electrical or thermal insulative specifications, as well as ways to create optimum physical properties. The roles of metal/polymer composite conduits and pipes are bound to multiply because of the versatility of the pultrusion processes.

A light-weight drive shaft of glass fiber and graphite was pultruded over an aluminum tube by Morrison Molded Fiber Glass (MMFG) Company in an application of aerospace technology to a high volume automotive production. This unique one-piece drive-shaft of a metal/polymer composite made its debut on General Motors 1988 pick-up trucks. The shaft is 4 inches in diameter, the wall thickness of aluminum .083 inch, and the fiber reinforced polymer .065 inch.[32]

Pultruded profiles are not always concentric and may utilize profiled shapes similar to I-beams or, in recent examples, the profile of an airfoil. Weight saving considerations are involved, and optimum selections of metal for strength and stiffness, accompanied by the surfacing, and the application of corrosion resistant layers of fiber reinforced polymers are factors present in new automotive applications. Heavy duty truck cabs are being designed with these objectives in mind.

Metal/polymer composites continue to proliferate in the automotive industry. The application of plastics in the making of different parts of automobiles was documented by Toray Research Center of Japan.[54] These applications suggest potential uses for many materials. Not all will be composites, but an examination of Table 6-3 will stimulate some ideas.

An innovative development of small automotive engines has been undertaken by Holzberg.[56] The engine, called a polimotor, utilizes glass fiber reinforced phenolic molding compounds as engine components. Iron cylinder liners, aluminum cylinder head inserts and steel internal reciprocating parts, coupled with a molded phenolic cylinder block, head, oil pan, oil and water pumps and intake manifold comprise a formidable metal/polymer composite.

Table 6-3. Application trends of plastics in making different parts of automobiles[54] (Toray Research Center)

6.1 Body parts

6.1.1 Body outer plates and exterior parts
  (1) Present status of plastic body outer plates
  (2) Development trends of plastic vertical outer plates
  (3) Development trends of plastic horizontal outer plates
  (4) Development trends of plastic body exterior parts
   A. Aero parts   B. Trim, Finisher, Grill, etc.   C. Outer handle
   D. Hubcap   E. Headlamp support panel
6.1.2 Bumper
  (1) Evolution of plastic bumpers and related regulations
  (2) Bumper system
   A. PP bumper   B. R-RIM urethane bumper
  (3) Required properties of plastic bumpers
  (4) Materials for plastic bumpers
  (5) Development trends of practical applications of automobile plastic bumpers
  (6) Future topics and prospects
   A. Topics   B. Prospects
6.1.3 Headlamps of automobiles
  (1) Headlamp lenses
   A. Application status of plastics for headlamps   B. Features of plastic headlamp lenses
   C. Topics and measures for plastic headlamp lenses   D. Future development
  (2) Reflectors plate for headlamp
  (3) Lamp cover
  (4) Tail lamp
  (5) Bulb socket of headlamp

6.2 Driving parts

6.2.1 Leaf spring
  (1) Features of FRP leaf spring
  (2) Application status of FRP leaf spring in Japan
  (3) Application status and future development trend of FRP leaf spring in General Motors Corp. USA
  (4) Composite spring of GKN
  (5) Topics of FRP spring
6.2.2 Pedals for automobile
6.2.3 Belts for automobile
6.2.4 Stator for automatic transmission

Table 6-3. (continued)

6.3 Engine parts
6.3.1 Plastic engine
6.3.2 Parts of general-purpose engine with integrated gear and camshaft
6.3.3 Automobile engine valves
6.3.4 Case of ignition device
6.3.5 Cylinder head cover
6.3.6 Rocker cover of automobile engine
6.3.7 Radiator tank
6.3.8 Surge tank
6.3.9 Gasoline tank
    (1) Market trends of gasoline tank
    (2) Characteristics and manufacturing methods of plastic gasoline tank
        A. Characteristics   B. Manufacturing methods
    (3) Regulation on the gasoline permeability of plastic gasoline tank
    (4) Methods for preventing permeation of gasoline
        A. Surface treatment   B. Multilayered structure
        C. Blending method of "Selar" RB barrier resin
    (5) Future development trend of plastic gasoline tank

6.4 Interior parts
6.4.1 Instrument panel
6.4.2 Console box
6.4.3 Roof
    (1) Structures of roof of automobiles
        A. Hanged roof   B. Bonded roof   C. Molded roof
    (2) Molded roof
        A. Resin coating molded roof
        B. Molded roof using polyolefin foam
6.4.4 Seat
    (1) Seat frames
    (2) Classification and molding methods of automobile seats
    (3) Seat made of heterohardness cushions
    (4) Polyester fiber pad materials for seats
    (5) Fabric springs for seats
    (6) Static-preventive seat materials
6.4.5 Steering wheel
6.4.6 Air bag
    (1) Explosive-ignition-method air bag system
    (2) Compressed-gas injection method
6.4.7 Skin materials for interior parts

Development and initial production are being carried out at Rogers Corporation. Weight savings of up to 50% over conventional engines are reported (Fig. 6-8).

POLYMER IMPREGNATION FOR STRUCTURAL PURPOSES

The impregnation and encapsulation of metal wires for structural purposes have important industrial applications in the preparation of polymer/metal composites. The physical attributes of the composites are due primarily to the fabricated metal components and the polymer impregnants enhance these qualities in the following manner:

1. "Dead" airspaces are eliminated in large insulative coil composites of electrical motors, generators, and transformers. This process minimizes

Figure 6-8. Molded phenolic automotive engines designed by M. Holtzberg are being produced by Rogers Corp. Iron liners are used inside the cylinders, forming unique metal/polymer composites. Courtesy of Rogers Corp. and M. Holtzberg.

the heat build up in electrical conductors which do not have the heat transfer capabilities made possible by contiguous insulation.
2. Noise levels and the movements of silicon steel laminations under alternating magnetic fields are minimized in electrical equipment.
3. Penetration of moisture and gases which corrode the metal components is prevented and physical integrity is assured.
4. The ability of an individual wire component to transfer physical loads such as shear stresses to its neighboring elements is structurally desirable.

Polymers for impregnation have preferably been low in viscosity, say 2000 centipoise or less, though higher application temperatures and preheated cores will lower the viscosities even more at time of application. To avoid the generation of gaseous by-products during cure, solvent free polymers such as epoxy resin systems (which contain cure agents) are used. Minimum shrinkage during cure is a prerequisite. To illustrate impregnation of metal assemblies for structural advantages, two examples are cited below.

1. Wire rope with improved wear and fatigue resistance was prepared by preheating the wire rope and injecting the plastic material, in a molten state, under heat and pressure in an extrusion machine. The interstices between the individual wires and strands are filled in this manner, while the lubricant initially present between the strands is substantially displaced. A smooth exterior can also be imparted to the wire rope. The impregnation of the wire rope with polypropylene more than doubled the number of bending cycles before failure.[33] Delmonte has also described the use of cast epoxy resins on wire cable terminations to enhance strength and durability.[34]
2. Coil of a large electrical generator is shown sectioned in Figure 6-9. The excellent performance of these units was ensured by the development of a solvent free, low viscosity anhydride cured epoxy resin capable of sustained performance at 200°C.[35] Industrial impregnation of motor coils have entailed the use of solvent varnishes for many years. Because of the presence of solvent (to facilitate penetration) complete void free impregnation, even with multiple dips, is not possible. The advantage of solventless impregnants is noted in the previous patent reference. In some practices, the coil is encased in a vacuum chamber and traces of volatiles are removed under vacuum. When this is followed by the resin impregnation, and the application of gas pressure, the best and most complete impregnation is assured.

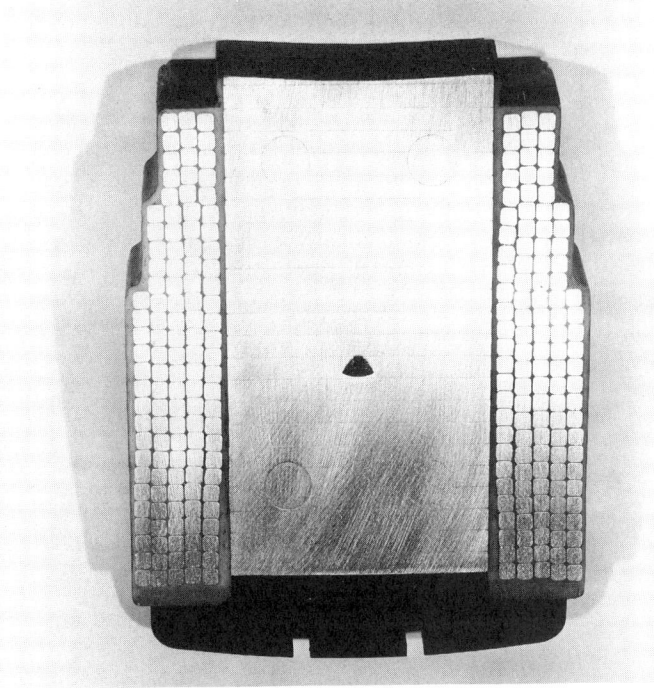

Figure 6-9. Large electrical impregnated coil with high heat resistant epoxy.[35]

LOW DENSITY METAL STRUCTURES

Low density and high porosity metal structures are produced in large volume by manufacturers of powdered metal parts. Generally, sintering at high temperatures, but below the melting point of the metal, produces a cohesive material which offers structural advantages. Conceptually, the porous, yet cohesive metal structure offers opportunities for metal/polymer composites. Close association of the metal particles contributes to electrical conductivity, as reviewed earlier in this book, and the sintered weldments between particles qualify the metal part as the matrix in the potential composite.

High porosity bronze composites function as self-lubricating bearings (see ASTM B438, B439, and B612). They are impregnated with petroleum based lubricants during manufacture. During service, frictional heat causes the lubricant to expand out of the pores and develop a low friction film. Copper-tin bronzes and copper-tin-lead bronzes are used for general purposes. As will be noted in examples listed further in this section, PTFE polymers, MoS_2 silicon oils, and graphite are also used as supplemental or primary lubricants.

156 METAL/POLYMER COMPOSITES

As an example of a porous aluminum casting, the units shown in Figure 6-10 depict acicular aluminum particles with a binder of a heat cured epoxy resin system. This unit was intended for use in a foundry, where the porosity of the casting facilitated its use as a drier support for sand cores passing through the oven. The appearance of the obvious porosity has prompted considerations of complete 100% polymer impregnation with electrical conductivity arising from contacts of the needle-like aluminum particles, thus producing better physical properties than may be obtained with fine aluminum particles.

The subject of a low average density and high strength will focus upon the many successful applications of honeycomb core construction. Figure 6-11 illustrates an open assembly of a simple flat panel and a core of *Nomex*, a duPont polyaramid. Surface plies could be made of reinforced plastic sheets or sheet metal. It is apparent that many different possibilities of metal/polymer composites are available. Earlier in this chapter, attention was directed to aluminum coatings on honeycomb walls for better resistance to water. Regarding honeycomb structures, the point can also be made that the greater the depth of the core, the lower the density that can be achieved, with very little sacrifice

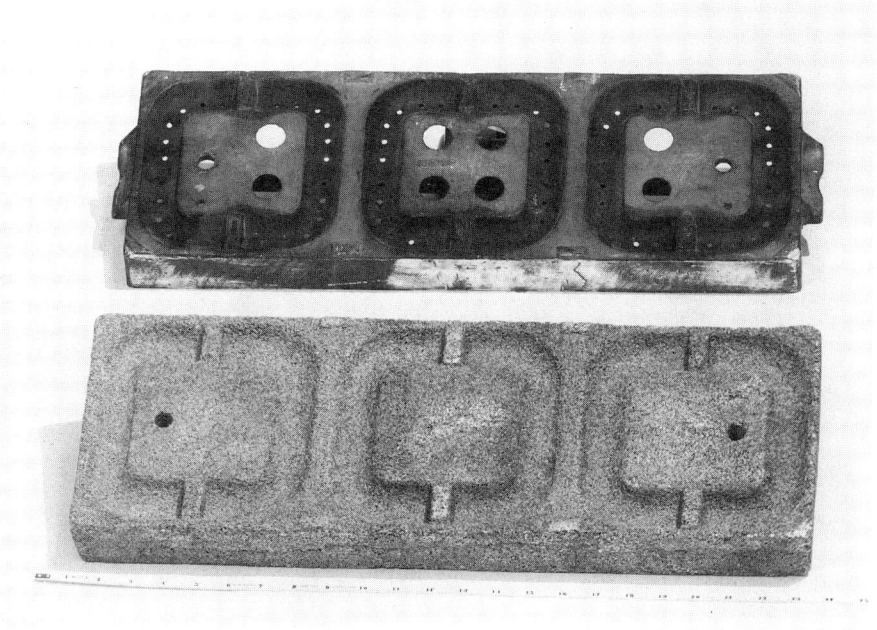

Figure 6-10. Porous aluminum casting. Courtesy of Furane division of Ciba Geigy.

METAL/POLYMER STRUCTURAL COMPOSITES   157

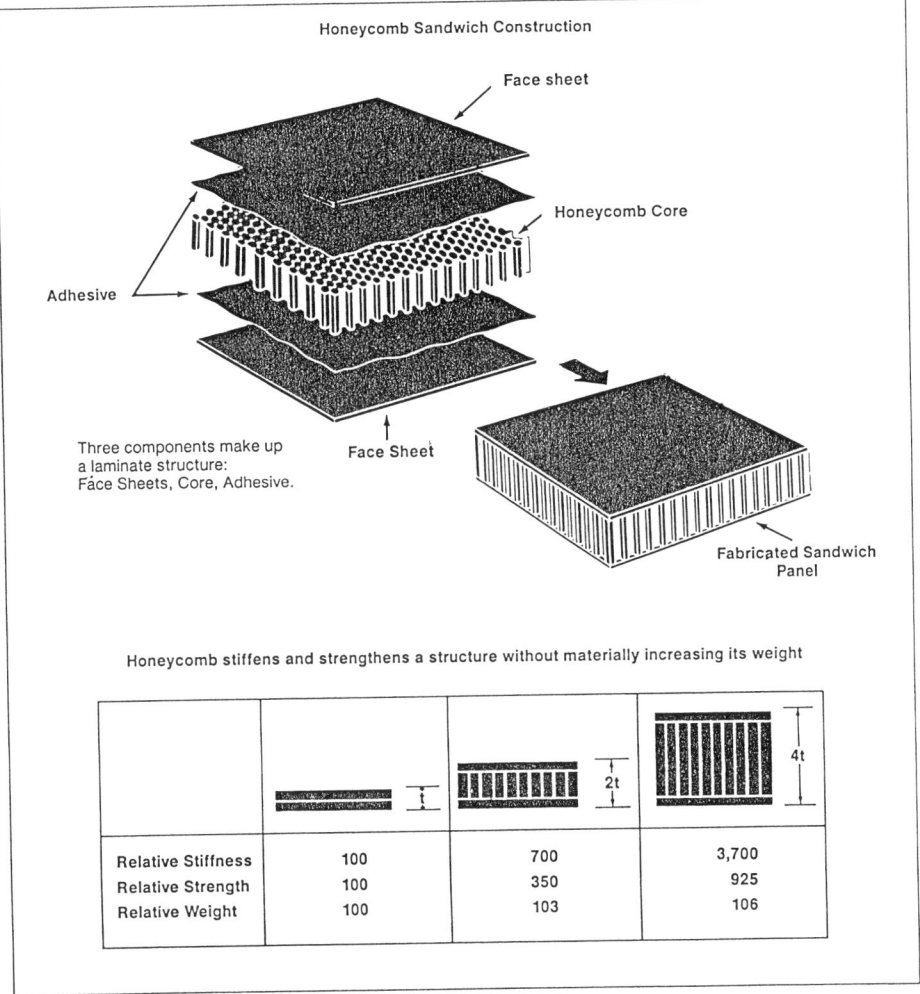

Figure 6–11. Honeycomb core construction. Courtesy of E.I. duPont de Nemours.

in strength. New opportunities continue to arise for these metal/polymer composites for use in light-weight structures.

The following paragraphs are presented as a small sampling of contemporary patents in porous metal structures, with emphasis on those used as bearings. With the availability of new high temperature polymers such as polyphenylene sulfide, polyether ether ketones (PEEK), polyimides, polyamide-imides, etc., and the use of high temperatures to achieve flowability, injection molding into porous sintered metals is possible in a closed mold. Unique metal/polymer composites can be developed and density variation achieved by the prior control of the sintering process. Synergistic balances which developed between the metal particle networks and the encapsulating polymer system will be observed.

- Sintered compact powder mixtures are prepared from C-fibers (7 to 8 μm) (5%), P-bronze (10%), and tetra-fluoroethylene powder (85%). The composite is pressed at 300 to 500 kg/cm^2 and sintered at 380°C for service up to 260°C. These compacts have functioned as self-lubricating bearings.[36]
- Low density porous metallic spheres for self-lubricating bearings; nickel plated for alkaline batteries, fuel cell electrodes, clutches, etc.[37]
- The effect of sintering temperatures was determined on porous filters prepared from carbonyl nickel powder (.1 to 4.2 μm). At 200°C to 500°C, segregation prevailed over volume shrinkage. At 900°C, the volume shrinkage predominated.[38]
- Porous materials suitable for increased heat and corrosion resistance were formed from organic molecules and metals into a three-dimensional network. Polyurethane foam was spray coated with Fe-Cr-Al powders and talc. The composite was fired for two hours at 800°C in an inert gas. The product was suitable for auto mufflers.[39]
- The effect of film thickness on the friction properties of thin polymer films has been evaluated.[40] Films of siloxane polymers .01 to 0.5 μm are present on computer memory disks. Poly (chloro-trifluoroethylene) was also examined.
- Mild steel with a porous layer of sintered bronze is impregnated with a polymer/metal mixture of polyvinyl difluoride powder (less than 20 μm), lead powder, and polythiophenyl-clay mixture. The composite was sintered at 675°C for one minute.[41]
- Short fiber preforms for composites are being prepared by Toyota Motor Company. Aluminum oxide and silicon dioxide fibers 2 to 3 μm in diameter and 2 mm in length are imbedded in a 10% colloidal aluminum oxide binder. A cylindrical preform 95 mm in diameter and 20 mm thick is dried out at 100°C for 30 minutes. It is then pressed with a spherical indenter leaving 21% volume fiber in the center and 7% volume fiber at the edge. This preform is infiltrated with molten aluminum alloy at 1000 kg/cm^2. The outer edges are easily machined for ring grooves.[46]

METAL MATRICES

Developments in metal polymer composites must take cognizance of the significant research, now in progress, to develop high strength and high temperature resistance from metal/ceramic fiber composites. The United States, West Germany and Japan have been the most active in these areas, with major applications planned for high speed aircraft, superconductors as applied to train levitation, and space vehicles. Silicon carbides and aluminum oxides, both as fine whiskers and continuous fibers, are being carefully considered. In commenting on the importance of weight reduction, Dr. Saito pointed out that while savings of only $2.50 per kg may be expected in surface vehicles, $120 per kg could be expected in conventional aircraft; cost savings of up to $10,000 per kg are anticipated in space vehicles.[42]

Examples of developments in metal matrices appear in the following paragraphs. They show the influence of light weight, high strength, and high temperature performance requirements.

Expanded honeycomb cores have been manufactured from titanium sheets. The formation of sandwich structures and diffusion bonding to face sheets that occurs in a single process has been developed by Douglas Aircraft.[43] Experimental parts for aircraft have been fabricated from titanium 6Al-4V sheets.

Graphite/magnesium castings have been developed by Aerospace Corp. The coating process makes graphite fibers wettable by magnesium in air enabling processing flexibility that had not been possible before. The graphite fibers are placed at strategic positions in a steel encased plastic mold before the magnesium is cast.[44]

Reports by Toyota Motor Co. on carbon fiber reinforced magnesium alloys indicated that magnesium alloyed with zinc (2 to 8%), zirconium (less than 2%), and aluminum (less than 1%) improved composite strength, and that interfacial reaction to form carbides is prevented.[45]

Silicon carbide (SiC) whisker reinforced aluminum composites are prepared by infiltration with molten aluminum (6061)—SiC volume was 17% and modulus of elasticity is greater than 100 GPa.[47]

Molten aluminum alloy at 780°C is used to infiltrate a bundle of long SiC fibers with vacuum assist of 10^{-4} torr. The product was pressed at 500°C for 30 seconds. The modulus of elasticity was 1.6×10^4 kg/mm^2.[48]

A polymer/metal moldable composite was prepared from powdered aluminum alloy—(less than 200 μm size) (25 pbw); short graphite fiber T-300 (25 pbw); and 1 pbw of powdered phenolic resin (less than 150 μm). The composite material was heated in a mold at 750°C for 15 minutes at 500 kg/cm^2.[49] Transverse strength measured 37 kg/mm^2.

Oriented fiber aggregates, such as stainless steel filaments (25 μm), are obtained by passing the filaments through a zinc-5% Al alloy bath at 420°C.

Ultrasound at 18 KHz and 300 watts aids the infiltration process. The composite wire with its metal matrix has a tensile strength of 70 kg/mm^2.[50]

Ube Industries in Japan has developed a silicon-titanium-copper fiber for reinforcing metal matrices. Its chemical stability in molten aluminum and its thermal stability up to 1300°C make it a viable candidate for high temperature structures.[51]

Nicalon continuous-fiber SiC/Al composite wire being produced by Nippon Carbon Co. of Japan is being distributed in the United States by Dow Chemical Co. Fiber volume of 45%, wire diameter 600 μm, and tensile strength of 70 kg/mm^2 (140 KSI) characterize the wire. The wire lends itself to press forming of structural shapes.[52]

METAL/PLASTICS COMBINATIONS

Metal/polymer composites suggest an intimate, contiguous union of two materials with synergistic results. End products which perform better than would have been possible if only one material were used, are the expected results. There are combinations of metal products and plastic products which fulfill their individual roles, with the products in close proximity to one another, though not necessarily with the close associations characterizing composites. These combinations are functionally very acceptable. A good example would be a portable, air circulating fan. The open frame enclosure for the rotating fan blade; the rotating fan blade; and the housing for the electric motor and its base are usually of a moldable plastic. The electric motor consists of insulated metallic electromagnetic components. The distinction between composites and combinations may be predicated upon the intimate contiguity of materials in composites as compared to the near proximity of the materials in the combinations.

There are instances when a material supplier has boasted of a physically and thermally improved polymer which may be used to replace metals. Such a material reflects an advance in polymer chemistry or materials processing. The materials engineer would be influenced by the selection of the best material for the purposes intended. In fact, efforts to place all of the design effort upon the merits of a new ubiquitous material, may not be as feasible as acknowledgment of the more obvious advantages of a standard material in combination with a compatible material of a different "species" of material.

Advances in materials structures continue to surge to the forefront. There was a time when the structure of an internal combustion engine was solely the domain of steel and other metal alloys. The availability of new "ceramic" lined cylinders have introduced concepts which can offer new, improved engines.

Metal matrices for nonmetallic fibers and ceramic fiber reinforced metallic wires have yielded structural materials of remarkable strength and durability. Nonmetallic fibers of remarkable strength have become available in recent years. These developments are not without processing problems, but reinforced metal matrices and electroconductive polymers, are seeking their niches in the growing roles of metal/polymer composites!

REFERENCES

1. Anon., *Plastic Trends*, 4: 25, (Dec. 1987).
2. Delmonte, J., *Aero Digest*, 60: 52, (Jan. 1950).
3. Delmonte, J., *Materials and Methods*, 40:93, (Aug. 1954).
4. Delmonte, J., *Tool Engineer*, 37: 84, (July 1956).
5. Anon. *Technology Review (M.I.T.)*, 90: 11, (Nov./Dec. 1987).
6. *ASTM Standardization News*, Spec. B-750-'85, (Jan. 1986).
7. East, W., *Materials Engineering*, 104: 28, (Oct. 1987).
8. Shell Chemical, U.S. 3,595,942
9. Lutz, R. and Gergen, W., (Shell Oil) U.S. 4,601,941, (July 22, 1986).
10. Bucci, R., et al., *33d SAMPE Proceedings* p. 1287, (March 1988) and *Bulletin of Aluminum Co. of American on "Arall"*
11. Brachman, A., (Bethlehem Steel Corp.) U.S. 4,229,504, (Oct. 21, 1980).
12. Hedrick, R., et al., (Monsanto Chemical) U.S. 4,369,222, (Jan. 18, 1983).
13. Buzet, et al., U.S. 3,382,136, (1964) and Zunich, M. J., U.S. 3,352,742, (1964).
14. C.A. 106-157-707R, Sigre, G.M.B.H., Ger. Off. DE 3,527,912 (Feb. 12, 1987).
15. C.A. 106(22) 177590A, Hitachi J.P. 62 09.017, (Feb. 26, 1987).
16. *Tech. Prod. Info.*, Westinghouse Copper Clad Laminate System 90, Grade 65M90, (1987).
17. C.A. 106-160994N, Bridgestone Corp., J.P. 61,222,711, (Oct. 3, 1986).
18. C.A. 106-139435N, Sumitomo Electric Industries J.P. 61,136,561, (June 24, 1986).
19. Kadash, M., et al., *Plastics Engineering* 41(12), (1985).
20. C.A. 106(20)-158-103C, Mueller, K., Ger. Off. DE 3618.478, (Jan. 29, 1987).
21. Newman, R., et al. to Dow Chemical Co. U.S. 4,313,996, (Feb. 2, 1982).
22. T. Inui (Toyo Kohan Ltd.), U.S. 4,614,691, (Sept. 30, 1986).
23. Snyder, G. and Stewart R. to U.S. Steel Corp., U.S. 4,204,022, (May 20, 1980).
24. C.A. 100-157907G, Nishin Steel Company, Can. Pat. 1,158,146, (June 27, 1980).
25. C.A. 101-132582, J.P. 59,10,306, (March 8, 1984).
26. C.A. 105-154,346 R. Showa Denko, J.P. 61,110,546, (May 28, 1986).
27. American Cyanamid, *Insights Bulletin*, (Dec. 1986).
28. C.A. 106(22) 177590a, Hitachi Chemical Co. J.P. 62,09,017, (Feb. 26, 1987).
29. Delmonte, J., *Technology of Carbon/Graphite Fiber Composites*, New York: Van Nostrand Reinhold Co., (1981).
30. Morris, E. E., in *Composites for Extreme Environments* ASTM-STP-768,—N.R. Adsit, Ed., A.S.T.M., pp. 95—109, (1982).
31. C.A. 105(18) 154255K, Guildford (U.K.), Composites Vol. 17, p. 217, (November 2, 1985).
32. MMFG, Application Profile, 129, Bristol, VA, (1988).
33. L. Smyth, U.S. 4,609,515, (Sept. 2, 1986).
34. Delmonte, J., *Wire and Wire Products*, 34: 1092, (Sept. 1959).
35. Delmonte, J. and Hirosawa, F., U.S. 2,967,843, (Jan. 10, 1961).
36. C.A. 100-125535T, J.P. 58,167,696, (Oct. 3, 1983).

37. C.A. 104(20) 172818E, G. Abbaschian, U.S. 4,565,571, (Jan. 21, 1980).
38. C.A. 100(24)196,202T, Poroshk Metall. Kiev, (June 1984).
39. C.A. 100(24)196,474H, Mitsubishi, J.P. 58,204,859, (1984).
40. C.A. 106(12)85849p, Trerill Iznos (USSR), Vol. 7, (965) (March 1986).
41. C.A. 106(18), Garlock Bearings, (May 14, 1987).
42. "Conference on Future of Advanced Materials" *C. and E. News,* p. 7, (June 15, 1987).
43. "Current Highlights," *Metal Matrix Composites,* Washington, D.C., 5 (1), (1985).
44. "Current Highlights," *Metal Matrix Composites,* Washington, D.C., 7 (1), (1987).
45. C.A. 104-134,449c, Toyota Motor Co., (1986).
46. C.A. 105-177,102X, Toyota Motor Co., Eur. Pat. 189,508, (August 8, 1986).
47. C.A. 106-88636C, Mitsubishi, Materials Sci. Tech. 3 (1): 57, (1987).
48. C.A. 106-88881d, Honda Motor Co., J.P. 61,207,534 (Sept. 13, 1986).
49. C.A. 106 88935E, Toray Industries, J.P. 61,195,939, (Feb. 27, 1986).
50. C.A. 106(18)142396Q, Mitsubishi, J.P. 61,210,136, (Sept. 18, 1986).
51. "Current Highlights," *Metal Matrix Composites,* Washington, D.C., 4 (2), (June 1984).
52. "Current Highlights," *Metal Matrix Composites,* Washington, D.C., 4 (4), (Dec. 1984).
53. Y.W. Mai, *Jl of Material Science Letters,* 7: 581, (June 1988).
54. New Developments in Automobile Materials for the 90's, Tokyo: Toray Research Center-Inc., (1988).
55. Meyers, R., *Handbook of Pultrusion Technology,* Chapman-Hall Publ. (1985).
56. Brooke, L., *Automotive Industries,* (May 1988).
57. ASM International, Ted Reinhart, Technical Chairman, *Engineered Materials Handbook,* Vol. 1, (1987).

7

RADIATION SHIELDING BY METAL/POLYMER COMPOSITES

The use of metal/polymer composites is well established in structural and electrical applications, and in innumerable electronic devices. Earlier discussions in Chapter 4 on electroconductivity, in Chapter 5 on coatings, and in Chapter 6 on structures, have examined many of the trends taking place in developments of metal/polymer composites. These include a large number of materials and processing methods, some of which are definitely experimental, though harbingers of things to come in the near future. In this chapter, attention will be directed to large area applications of metal films, as well as polymer/metal molded composites. Particular emphasis is placed upon the electrical applications which include EMI and RFI shielding and the intermediate electrostatic control composites. Because continuous metal films applied upon polymer films bear physical similarities to electrical shielding materials, important applications in food packaging and product applications are included in a separate section. Brief discussions are also made of coaxial cables and of nuclear radiation shielding.

The subject of microapplications of polymer/metal composites is examined in Chapter 9. Micro prefixes refer to dimensions in the micrometer range or less. There has also been a necessity to make references to nano (10^{-9}) units, particularly among semiconductors. Emphasis is placed upon materials and the very compelling roles placed on micro metal/polymer composites which require ingenious concepts and close relationships of the metals and nonmetals.

EMI (ELECTROMAGNETIC INTERFERENCES) AND RFI (RADIO FREQUENCY INTERFERENCES)

EMI/RFI shielding has been growing in importance as the number of electronic devices in the home and in industry has greatly increased. Electronic "noise" pollution has arisen from home entertainment and communications equipment, the early CB radios, home computers, and video games linked to television, automobile ignitions, electronic fuel-injection systems, etc. Stray EMI signals can also affect critical electronic equipment, such as computers, medical instruments, navigation equipment and process control equipment. House and automotive interference sources are usually below 100 MHz, while digital computers exhibit interference in a higher range, up to 1 GHz.[1,2]

Regions associated with high frequency radiation are tabulated in Table 7-1. It will be noted that the EMI is of concern within a narrow region, 10^5 to 10^9 Hz, encompassing FM, AM, and TV broadcast frequencies. The Federal Communications Commission (FCC) became active in the early 1980s in establishing regulations which would reduce the amount of interference (static) issuing from thousands of electrical and electronic sources. In earlier years much equipment had been housed in sheet metal containers and hence had inherent EMI shielding from the metal casings. Recently molded plastic cases, because of their light weight, attractive appearance, and low cost, supplanted the metals. However the plastic containers lack shielding unless metal powders or metal filaments have been compounded into the polymer.

Concern with the interference of radio frequency communications first arose in the 1930s when the ignition systems of piston driven aircraft engines and automotive engines were cited for their R.F. noise. At that time, the author described wire-braided shields developed to confine the extraneous signals arising from ignition cables.[3]

The FCC regulations (FCC Docket 20780) now require that all electrical/electronic products manufactured after 1983, not exceed specified EMI/RFI levels of interference. These regulations are concerned with emission from assembled electronic devices rather than shielding values based on flat panel

Table 7-1. Regions of High Frequency Radiation

REGION	WAVELENGTH ANGSTROMS*	APPROXIMATE FREQUENCY RANGE (Hz)
Radio, TV, AM	10^{15} to 10^{10}	10^5 to 10^9 Hz
Microwave	10^{10} to 10^8	10^9 to 10^{12} Hz
Far infrared**	10^7 to 10^4	10^{12} to 10^{15} Hz
Visible	400 to 750 nanometers	10^{14} to 10^{15} Hz
Ultraviolet	10 to 400 nanometers	10^{15} to 10^{17} Hz
X-Ray	.01 to 10 nanometers	10^{14} to 10^{22} Hz
Gamma rays	10^{-3} to 10^{-4} Nanometers	10^{22} to 10^{23} Hz

* 1 Angstrom = 10^{-10} meters
1 Nanometer = 10^{-9} meters
1 Micrometer = 10^{-6} meters
10^6 Hz = 1 Mega Hz (MHz)
10^9 Hz = 1 Giga Hz (GHz)
ASTM-E-135 Vol. 3.06 lists ranges of wavelengths
*For good thermal imaging through the atmosphere—8000 nanometers to 12000 nanometer wavelengths are recommended
Radar-typical-3×10^8 angstroms

material testing. In this manner, protection must be devised for all apertures and connections of the devices.

The spectrum of high frequencies shown in Table 7-2 indicates the radio frequencies allocated in the hundreds of kilohertz, which follow the Earth's curvature to some extent. These wavelengths do not transmit high speed data, or high quality speech or music. The higher frequencies in tens or hundreds of megahertz carry wider bandwidth signals, and hence more information, since they resort to frequency modulation instead of amplitude modulation. These high frequency signals travel in straight lines. Signals in the UHF band, from 300 to 3,000 megahertz (3 gigahertz), also travel in straight lines, and hence are appropriate for the use of satellites and satellite receiving dishes.

Super high frequencies (SHF) from 3 to 30 gigahertz extend the range of communications, and as reliable circuitry for processing UHF and SHF become available, more opportunities arise for their use. Aircraft may receive signals from satellites as aids to navigation or may bounce signals off satellites for communication. Deep space probes condense much data into a short time interval for abbreviated high frequency transmission.[42]

It is obvious that shielding needs arise not only in consumer oriented products, but also in very high frequency communications, and the roles of metal/polymer composites will become more sophisticated to fulfill these requirements. The chapter on semiconductors identifies some of the roles that have been established for metal/polymer combinations. With a new generation of solid-state devices (of composite metals/polymers/ceramics), optical circuitry will take over at the very high frequencies and permit exploitation of new areas of the electromagnetic spectrum.

Electromagnetic radiation has two components, the electric field, E, and the magnetic field, H. The ratio of the electric field to the magnetic field is called wave impedance and is measured in ohms. At a short distance from the radiating source, near-field conditions exist; and at long distances, far-field conditions prevail. Wave impedance becomes a constant, determined by the nature

Table 7-2 FCC Limits for Electromagnetic and Radio Frequency Interference

	FIELD STRENGTH (MICROVOLTS PER METER)	
FREQUENCY RANGE (MHz)	CLASS A	CLASS B
30-88	30	100
88-216	50	150
216-1000	70	2000

A = Commercial and Industrial, measured at 30 meters
B = Residential, measured at 3 meters

of the propagating medium (for air and vacuum, the wave impedance is 377 ohms).[4] In the far-field region, the electric field *(E)* shielding effectiveness and magnetic field *(H)* are equal. The region between near and far field is called the transition region, and a transition point is defined as:

$$\text{Transition Point} = \text{Radiation Wavelength}/2\pi$$

The transition point *(TP)* can vary from a distance of 1.5 m from the source at 300 MGz to 5 cm at 1000 MHz. For near-field data, a dual chamber method is used and for far-field data, a transmission line is used.[5] For determining shielding effectiveness *(SE)* of planar materials and their comparative ranking, a sine wave signal generator is used with a 50 ohm input impedance. Measurements are made at 30 MHz, 100 MHz, 300 MHz and 1000 MHz. Calibrations are performed upon a composite of polyester film (.05 to .10 mm thick) coated with a gold film.

When radiated power, *P*, is incident on a plane of finite thickness (Fig. 7-1) its power is attenuated by reflection and absorption. When the coated metallic film is very thin (say 100 angstroms for a vacuum deposited metal), the absorption loss is small and can be ignored. Microwave attenuation due to coated metallic film on a plastic substrate is independent of both frequency in the far-field region, and surface resistance.[55]

$$P = P_R + P_A$$
P_R = Reflected Power
P_A = Power Absorbed Within the Plane

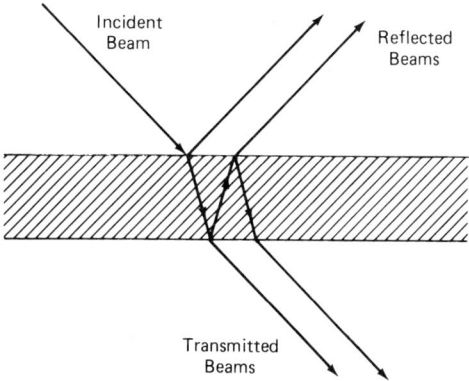

Figure 7-1. Microwave radiation reflects from the top and bottom surfaces of non-metallic materials.

A practical approach in determining the *SE* of a planar sample is to measure the strength of the signal transmitted at a fixed frequency (usually between 1 and 3 watts), with the sample in position, and with no sample in the test aperture.

$$SE = 10 \text{ Log} \frac{P_T}{P_I} \, (dB)$$

P_T = transmitted power through plane
P_I = power incident on the plane

The attenuation unit is the decibel (dB). A shield which transmits 1% of the radiation has an *SE* of 20 dB (re. ASTM-ES7-83), while a shield with *SE* of 40 dB will transmit only .01% of the incident radiation.

Shielding is realized by the combination of reflection losses and absorption losses. Absorption loss corresponds to the attenuation due to the skin effect at higher frequencies, and is dependent upon frequency, conductivity, and thickness of the barrier. Most EMI shields are made from metal films which will ensure high reflection losses. Household aluminum foils about .038 mm thick offer shielding effectiveness of 100 dB. To obtain shielding effectiveness at low frequency magnetic fields, metal films must be thick, or alternatively the composite may include highly permeable material such as *mumetal* which is an alloy of nickel, iron, and copper that is characterized by high magnetic permeability and low hysteresis losses.

Bigg has made a detailed study of the variables influencing the measurements of shielding effectiveness on composites.[6] As he points out, the transition region between semiconductive composites and electrically conductive composites covers a broad range of volume resistivities. It is likely that minor changes in formulation produce a large change in conductivity in this intermediate region, which will in turn affect the *SE* measurements if conductivity is marginally attained in manufacturing. The physical attributes of the devices used for measuring the EMI shielding affect the interpretation of the data. A metal structure must be welded along all seams and corners as outside radiation sources will interfere with signals received by the antenna. All entries through walls must be sealed with conductive sealants and pastes. Good electrical contacts are essential to the placement of the test panels. Should the composite under test be fabricated from molded or extruded material, the resin rich surface insulation must be lightly abraded to permit intimate contact with the conductive particles. At the higher frequencies for *SE* measurements, the coaxial cable and the insertion of composite samples at a test flange, require similar precautionary measures to prevent radiation from leaking in through openings.

Figure 7-2, adapted from Bigg[6] illustrates the theoretical relationship of

168 METAL/POLYMER COMPOSITES

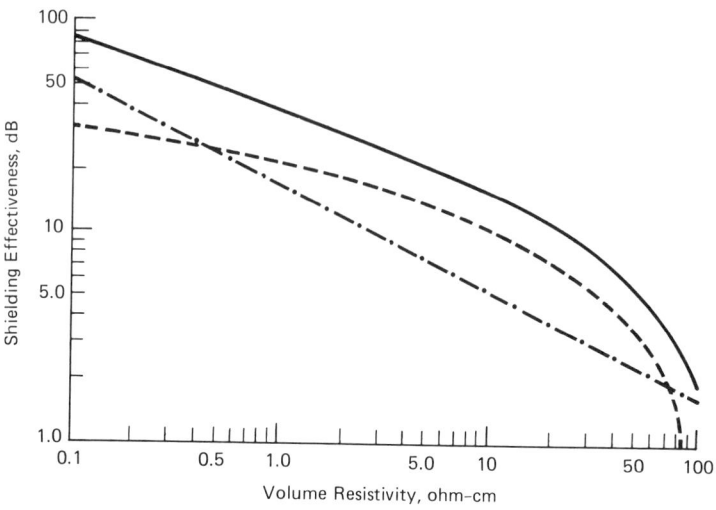

Figure 7-2. Plane wave shielding attenuation for a 3 mm thick material. Courtesy Bigg.[6]

volume resistivity and plane wave shielding attenuation for a 3 mm thick material at 1 GHz. Bigg also examines the volume resistivity relations of materials and the influence of the volume fraction of aluminum flakes in polypropylene, ABS, and polycarbonates, and the volume fractions of iron and aluminum powders, and fiber aspect ratios. These data demonstrate the rapid change in volume resistivity with small changes in volume concentration, in the regions between semiconductors and good conductors (see Chapter 4). Also relevant to this study is the relationship between surface resistance and light transmission through thin copper and silver films[49] (Fig. 7-3).

Recent trends in the adaptation of metal/polymer composites to the reduction of EMI/RFI emissions cover a wide spectrum of materials. Emphasis has not only been on the fulfillment of FCC regulatory levels, but also on cost effective applications, which influence large volume users. The percentage of the shielding market held by different materials and application techniques is given in the table below, which was prepared by Business Communications Co., Inc.[7] Also to be considered in addition to the data below, are the applications to magnetic tapes (conductive coatings), recording discs, and the reflective panels of new high-rise buildings (vacuum metallizing).

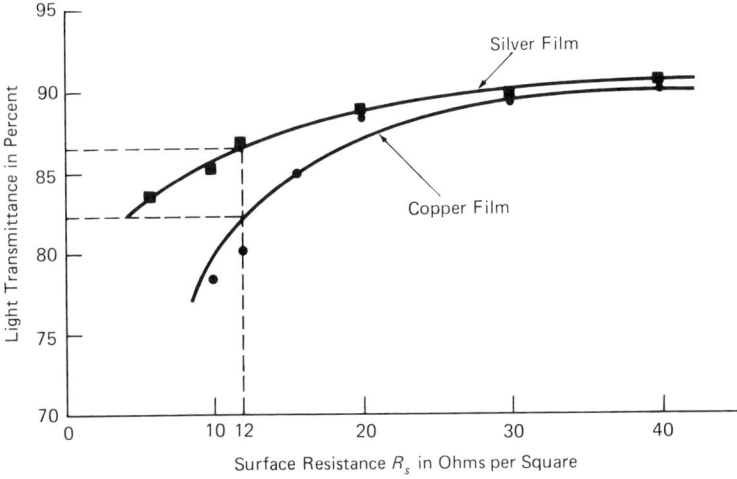

Figure 7-3. Light transmission vs. surface resistance of copper and silver films. Courtesy Liao.[49]

1982		1986
	Conductive coatings	
89%	including Zinc Arc Spray	79%
4%	Vacuum metallization	2%
	Electroless plating	8%
	Conductive plastics	0.5%
7%	All others	10.5%

During the years 1986–1987, the use of zinc arc spray coatings increased dramatically. They have been applied to entire rooms, buildings, and ships as protection against spy network penetration of areas with sensitive electronic equipment. Improved zinc spray coatings (Chapter 5) as with 5% Al-Misch metal, has helped reduce the quantity needed for effective spraying. Material costs of $6.00 per square meter produce a coating thickness of .003 to .010 inch. These large applications of zinc arc sprays have supplanted the techniques used in 1982, when many cabinets for computers and radio and TV equipment were treated.[8] The high power radar systems of a modern airport can cause computers to fail unless the computers are screened by protective composites shielding. Electronic switching in automobiles and home appliances are subject to these problems. Europe has lagged behind the United States in adopting protective measures.[9]

The development of electroconductivity in polymers was discussed in earlier chapters which reviewed the addition of conductive metallic fillers to insulative polymers; the use of metallic fibers, including metal coated graphite and thermoplastic fibers; and the many procedures for the metal coating of plastics, as by electroplating and ion implantation; spraying, and vacuum deposition of metals. The marriages between metals and polymers have established new classes of metal/polymer composites. Shielding has been a prominent activity for many manufacturers. The many technical papers that have been published and the considerable attention given to these applications in trade and industrial shows, as well as the surge of patents that are now being issued in the United States and Japan show the necessity for emphasizing this subject.

Conformance of electronic computing devices with FCC Docket 20780 requires examination of the assembled units. It is necessary to test for device compliance, rather than determine the shielding values based on flat material testing. Correlation between end use testing and flat panel testing indicates conductive plastic with 30 to 50 dB of shielding can often replace conductive coatings, according to Gerteisen.[10] The point should be made, however, that testing of assembled devices has not been standardized and results may differ from one laboratory to another. The precaution of a well shielded room, including all outlets, is essential to prevent stray electromagnetic radiation. Measures are necessary for controlling electromagnetic interference from video display terminals. Germane to the subject matter of this book is the fact that extensive adaptations of metal/polymer composites are found necessary to achieve EMI/RFI shielding.

Among the ancillary materials used in preparing electronic devices for effective device shielding or testing rooms are the following:[11]

- Semi-resilient electrically conductive compounds are used to provide gaskets and seals to confine electromagnetic disturbances within their shield encasements. Flanges, O-rings, and gaskets are typical.
- Sealants such as copper or silver filled polymers will caulk and seal connections, and foam-in-place gaskets. The flexible gaskets should also be compounded with conductive fillers and/or conductive fibers.
- Adhesives which are electrically conductive are essential for the full bonding of metal components required in equipment meeting FCC regulations.
- Repair coatings of nickel and silver polymer composites for EMI/RFI shielding-circuit repair.
- Wire screens or metal honeycombs for apertures and openings.
- Wire braided harnesses for interconnects, etc.

Electrically conductive fillers and reinforcement (fibers) have been examined for some time, as can be gathered from the examples cited in Chapter 4. The cently, and will, in time, supplant the conductive fillers in some applications. Recognition of higher conductivity for EMI shielding appeared as metal/polymer composites began to replace carbon black fillers. Union Carbide found that sheet molding compounds with 3 grams/360 cm^2 of carbon/graphite fiber showed EMI shielding in the range of 25 to 40 dB at lower frequencies (27 to 120 MHz) compared to 75 dB for aluminum. At increasingly higher frequencies, to 1 GHz, the aluminum attenuation fell to 38 dB and the carbon/graphite fibers showed little change.[12]

Bulk molding compounds do not use the mats or veils of conductive material, but they do use conductive fillers and fibers, or pellets of conductive fibers extruded with compatible polymers. Compression molding of metallized glass fibers has proved more effective in EMI shielding rather than in injection molding where the molding pressures break down the high aspect ratio of brittle fibers. When the bleed off of static electricity is desired, the requirements of electrical conductivity are less, as will be noted in the next section of this chapter. Aluminum fibers, ribbons, and flakes are less brittle than metallized glass fibers. The considerable effort now being undertaken on EMI/RFI shielding is directed towards cost effective measures because of the large quantities of materials involved. Table 7–3 lists some of the characteristics of contemporary conductive materials, as developed for shielding purposes. This is followed by a listing of recent patent activity and reports on new material developments. As in earlier chapters, the examples will be drawn from establishments which have demonstrated commitments to metal/polymer composite technology. The sources of the data are from the references noted in the table. *SE* was measured with the aid of coaxial cable at the higher frequencies. A few volume resistivities are reported in examples of metal/polymer composites. The most effective shields for R.F. are made from materials with high electrical conductivity and low permeability. Electroless copper of 1.2 μm and a nickel overlay showed attenuations of 100 to 80 dB in the range of 30 MHz to 1,000 MHz.[4]

Conductive thermoplastics having high EMI shielding effectiveness have been prepared from several thermoplastics containing 25 to 40% (weight) aluminum metal flakes, 4 to 8 % of metal or metal coated fiber. Table 7–4 reproduced from Reference 13 demonstrates some of the formulations and the results that were obtained.

Significant improvements were obtained in the formulation by the addition of carbon black (XC-72) from Cabot Co.[14] The shielding effectiveness of the material was maintained with the added advantage of a greatly reduced surface resistivity on the addition of carbon. It should be remembered that surface conditions of molded parts are influenced by the molded surfaces, ''resin rich''

Table 7-3 Comparisons of Several Shielding Systems

EMI/RFI SHIELDING SYSTEM	CONDUCTIVE MEDIUM	SURFACE RESISTIVITY OHMS PER SQUARE	VOLUME RESISTIVITY OHM/CM	SHIELDING EFFECTIVENESS			REF.
				30 MHZ	100 MHZ	1 GHZ	
Zinc arc spray	Zinc	.020		106	92	98	(A)
Sprayed acrylic coating	Nickel	3.0	.01	35	47	57	(B)
	Silver	.004		67	93	97	
Panels of thermoplastic	5% NCG	20.0	0.5	32	30	34	(C)
Polycarbonate and	10% NCG	2.0	0.1	36	36	44	
polyetheramide	20% NCG	0.8	0.5	50	52	59	
(Nickel 0.4 μm on 7.0 μm graphite fiber)	5%	10^{13}	1.0	40	35	40	(D)(F)
Stainless steel fibers (7.3 μm) in polycarbonate	10%	10^2	0.5				
Aluminum flake (1 mm × 1.2 mm −30 μm thick) (from rapid solidification)	18 to 22% (volume)		1.0			35	(E)

A: M. Thorpe, *Plastic Engineering*, 38 (4), April 1982.
B: Al-Technology Inc., *Technical Bulletin*, 1986.
C: A. Luxon and M. Murthy. *SPE Antec*, page 233, 1986. R. Evans et al., *SAMPE Quarterly*, 17: 18, July 1986.
D: S. Gerteisen, *Northcon/85*, October 1985. R. Tolokan and J. Noblo, Plastics Engineering. 41: 31, August 1985.
E: A. Holbrook, *Intl. Journal Powder Met.*, 22 (1): 40, 1986.
F: S. Kidd-*SPI-42d Annual Conference-paper 25F*. Feb. 2-6, 1987. Composites Institute of S.P.I.

Table 7-4. Metal Polymer Compositions Evaluated[13]

MATERIAL	PARTS BY WEIGHT		
Aluminum flake	36	36	36
Aluminum fiber	4	4	4
Polycarbonate	60		
Poly (ethylene terephthalate)		60	
Nylon 6-6			60
dB shielding effectiveness at 0.5–1 GHz	59–46	58–42	49–38

thicknesses on the surface, would affect test results. Processing difficulties, particularly involving thermal blending, will also influence the results and reduce the aspect ratios of the fibers and the metal flake additions.

Because of the relatively recent appearance of the EMI/RFI shielded materials, published data on the stability of performance under environmental extremes, temperature extremes, and physical operating stresses are not always available. The long time effects will, eventually, aid in the making of decisions for choosing optimum metal/polymer composites.

High energy Compton photoelectrons are released from metal surfaces after exposure to electromagnetic pulses or x-rays. These high energy electrons, when emitted from high atomic number elements such as gold, are reported to cause damage to sensitive electronic systems.[15] Thin 25 μm films of an electrophoretically deposited styrene-acrylate coating on the metal surface will suppress the internal electromagnetic pulse effect. Dielectric coatings are low electron emission materials.

In addition to EMI/RFI shielding qualities, there are other advantages which may occur when using conductive fillers. For example, aluminum flakes which are present in molded plastic parts constitute an advantageous heat sink which aids in dissipation of heat generated in electronic components (Fig. 7-4). The light colored streaks are cross-sections of the aluminum flakes. Also, nickel coated graphite fibers influence the thermal expansion coefficients of the composite into which they have been introduced, as does the thickness of the nickel plate which may have been electro-deposited on the graphite fiber. In general, increasing volume percentages of metal fibers or metal coated fibers will increase the impact strength of the composite.

Metal powders, metal fibers, metal-coated fibers, and very fine metal fibers from electroless copper and nickel (Chapter 5) are present in metal/polymer composites for EMI/RFI shielding. There are other forms which should be

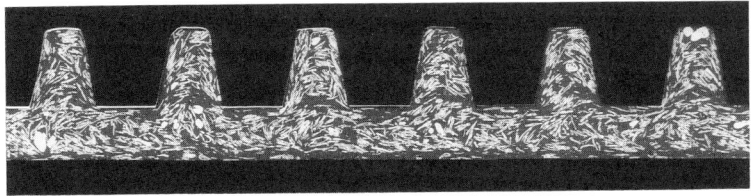

Figure 7-4. Aluminum flakes in a molding compound contribute to EMI shielding and function as a heatsink. Courtesy of Transmet Corp.

noted, including aluminum honeycomb which has been used to shield heating and ventilation openings from the passage of RFI.[16]

COMPARISON OF TECHNOLOGIES FOR SHIELDING

The prevalence of electronic "noise" continues to grow and places more emphasis on the shielding requirements. Power tools, welders, and malfunctions in home entertainment and communications equipment, even paging systems, have shown up on EKG charts. Office business machines, microprocessor controls, and electric appliances emit EMI at a number of frequencies and in turn are susceptible to the presence of external disturbing machinery. Aircraft control towers have experienced the effects of disturbing electronic frequencies. The needs for shielding by metal/polymer composites are growing.

Zinc metal spray gives a continuous metal coating, though complex shapes are hard to coat uniformly. Organic coatings of metal powders, such as finely divided silver or nickel are easy to apply, though they are easily influenced by the environment. Long term adhesion and the application of uniform thickness may offer difficulties. When thin, though fragile coatings, such as those obtained in vacuum metallization are used, the metal film must be protected by polymer coatings. Electroless metal coatings of nickel or copper have exhibited good shielding performance when measured against the EMI/RFI frequencies. An electroless coating followed by electroplating, which adds more expense, gives good reliability. Finally the metal filled polymer molded parts have minimum processing requirements and possess the physical qualities necessary for permanence. They are to be preferred for large productions of small molded parts requiring high shielding effectiveness.

Conductive composites for shielding are being successfully prepared from moldable thermoplastics incorporating conductive metal particles, flakes, carbon black, and fibers. Material suppliers such as Wilson-Fiberfil, Ferro Corp., Premix Inc., LNP Inc., and others have been identified with these products. There is no exact relationship between electrical conductivity and shielding effectiveness, as the compounding procedures may introduce variables; and in application by the consumer, the inadequacy of protection of all apertures, which may admit radiation, may negate the performance of a conductive compound. Test methods, some of which have been described in this chapter, are being standardized and progress is being made.

Gerteisen[43] has made a convenient comparison of the effectiveness of shielding of several conductive fibers of carbon, nickel coated graphite fibers, and stainless fibers (all about 7 to 8 μm in diameter). Figure 7–5 depicts the results of tests per ASTM-ES-7-83. The electroplated PAN (polyacrylonitrile) carbon fibers had 0.5 μm of nickel on the surface. The shielding effectiveness (SE) in

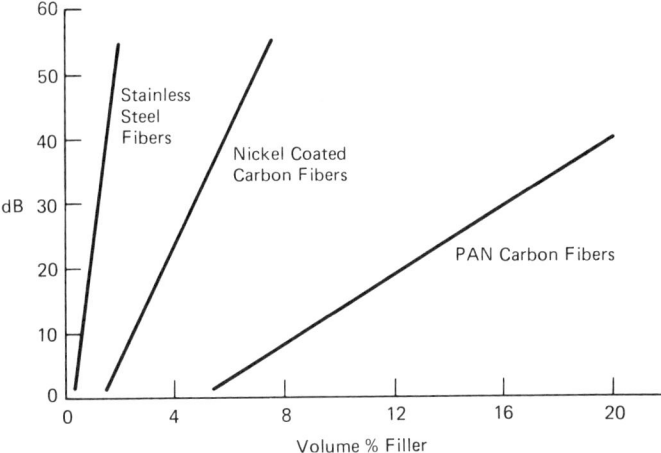

Figure 7-5. Comparison of shielding effectiveness with several conductive fibers. Courtesy of Wilson-Fiberfil.[43]

Table 7-5. Typical Properties of Molding Materials For EMI Shielding and Electrostatic Discharge *

ASTM	ELECTROSTATIC DISCHARGE	EMI SHIELDING
D638—Tensile strength (psi)	8,000	7,000
D790—Flexural strength (psi)	20,000	18,000
D648—Heat distortion °F at 264 psi	400°F	400°F
D257—Surface resistance ohms/square	10^4 to 10^5	10^2
D275—Volume resistivity ohm-cm	10^3 to 10^4	5 to 10
C177—Thermal Conductivity Btu/ft²/hr (°F/in.)	30 to 38	40 to 45

* From Premix Inc., *Technical Data Bulletin*, North Kingsville, Ohio, 1985.

this figure is an average value, as *SE* is usually higher at 1,000 MHz frequency. Table 7-5 lists physical and electrical properties of typical molding compounds for anti-static control and EMI shielding.

There have been many patents on the development of EMI/RFI shielding formulations. Some of the more current developments are summarized in the paragraphs below.

Synthetic resins are injection molded with a low melting alloy consisting of 48% Bi, 28.5% Pb, 9% Sb, and 14.5% Sn. Electromagnetic wave insulation is provided by the injection molded cabinets.[17]

Acrylic ester resin solution is mixed with colloidal silica, stearic acid, and globular copper particles (15 μm—0.5 m²/g). Excellent electromagnetic shielding properties are claimed.[18]

Mica flakes are electroless coated with copper and treated with a silane coupling agent (animopropyl triethoxysilane). Forty parts by weight of the coated mica flakes are melt kneaded with 60 parts by weight of polypropylene and formed into pellets. Injection molded sheets showed the following results, significant to EMI shielding:[19]

ATTENUATION (dB)	FREQUENCY
26	10 MHz
35	100 MHz
45	1 MHz

Electromagnetic shields are prepared from brass metal screens, alternately layered with nylon 6 containing 19% glass beads and 30% short glass fibers. This composition was molded into a box, which showed an attenuation of around 50 dB between 100 to 600 MHz.[20]

Chlorinated polyolefins are mixed with acrylate polymers and powdered silver, nickel, copper, chromium, aluminum, and zinc. Silver powder in the polymers is used to develop 25 μm films on 2 mm thick polypropylene cases for good electromagnetic shielding.[21]

Smoluk presented an analysis of various materials to meet the EMI/RFI shielding requirements.[22] He interpreted EMI shielding effectiveness *(SE)* in dB as follows:

S.E.	PRACTICAL VALUE
10–30	Minimum
30–60	Average
60–90	Above average

An electroless copper deposition of 25 μm was described as above average. It provided uniform coating regardless of part geometry, as the liquid solution penetrates holes and behind ribs.

Plastic components were given EMI/RFI shielding by sputter deposition from a magnetron. High conductivity copper and aluminum were studied. A laminate formed with a steel coating improved resistance to the marine environment.[23]

Electromagnetic shields for plastic housings were prepared from ABS copolymers, electroless plated on both sides with 2 μm of nickel. Polyurethane adhesive was applied on one side and covered with a polyethylene terephthalate (polyester) film, 30 μm. Attenuations of 30 to 35 dB at 20 MHz and 50 dB at 1000 MHz were reported.[24]

A mixture of dicyclopentadiene and diethyl aluminum chloride and monel metal filaments was injection molded to form a 3 mm electromagnetic shield. The attentuation was 40 to 60 dB over a range of 10 kHz to 10,000 MHz.[25]

Powdered copper filled acrylic coatings are used as linings inside of computer housings as EMI/RFI shields. These coatings are applied directly to the molded plastic cabinets and fulfill the prevailing regulatory requirements.[26]

For the protection of electronic devices from sound, radiation, and vibration, composites have been prepared by the addition of fine lead (Pb) flakes to a chlorinated butyl rubber matrix.[27]

Chlorinated polyolefins are mixed with acrylic powders and powdered metals, Ni, Cu, Cr, Al, Zn, and Ag to form a series of metal/polymer composites. In the silver-polyethylene-acrylic system, a 25 μm thick film was formed on the surface of a polypropylene case (2 μm thick wall). Good results in EMI shielding are reported.[28]

Stainless steel fibers (3 to 50 μm, 5 to 20%), glass fibers (5 μm, 4 to 30%), and ABS polymers are thermally blended, extruded and pelletized to form a metal/polymer composite for molding electromagnetic wave shields.[29]

Thin flakes of stainless steel (1 to 40 μm) and aspect ratios of 20 to 100 are treated with the reaction product of epoxy resin and silyl isocyanate. The compound for molding was prepared from 30% stainless steel flakes and 70% polyvinyl acetal resin. It is noteworthy that there was a modest gain in the physical properties of the molded metal/polymer compound.[45]

Stainless steel fibers are coated with polyvinyl copolymer, chopped to a 5 mm length and introduced into ABS molding compounds. With a stainless steel content of 43% (by weight), good EMI shielding is achieved. The product demonstrated 46 dB attenuation at 100 KHz and a volume resistivity of 0.3 ohm-cm.[46]

Stainless steel fibers have also been compounded with rubber modified polystyrene to form anti-static shields and EMI shields. Combinations of aluminum fibers were also included. The proportions used were described as 0.1 to 2% volume of stainless steel fibers (2 to 15 μm) and 0.1 to 10% volume of 50 μm Al fibers, both with high aspect ratios ($\approx$ 100).[47]

Borrowing from paper technology, International Paper Company has developed a light-weight random-fiber, non-woven mat with applications for shielding and conductivity.[48] With carbon fibers, weights as low as 6.3 g/m^2 have been prepared, and with polyaramid fibers 11.0 g/m^2 have been achieved. EMI shielding characteristics of carbon/mat laminates are shown in Figure 7-6 for different weights of mat. Carbon fiber lengths are produced from 2 to 5 cm. Nickel coated graphite fiber mat increases the weight to 25.5. g/m^2 and offers better EMI/RFI shielding. Vacuum aluminized mats of 50% carbon and 50% glass fibers, weighing 34 g/m^2 are also available. These products are used in business machine housings, antenna dishes, electrostatic surfacing, and pollution control exhaust stacks.

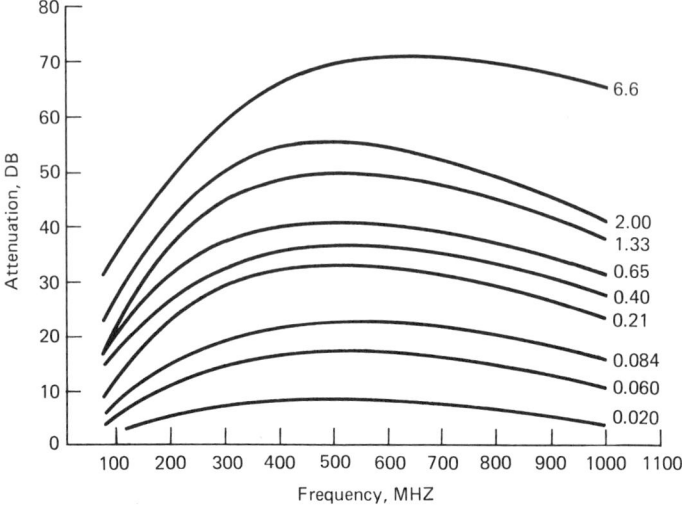

Figure 7-6. Shielding properties of carbon-mat laminates.[48]

CONTROL OF STATIC DISCHARGE

Electrostatic charges will accumulate on nonconductive surfaces which are rubbed together or which flow past one another and are separated. Materials will support a positive charge or a negative charge, and the electrostatic charge may create processing difficulties on films, paper products, and electronic devices. Electrostatic discharge (ESD) is defined as a transfer of electrostatic charges between bodies at different potentials, caused by direct contact, or induced by an electrostatic field. Difficulties have appeared in computers and delicate integrated circuits. Aluminum foil/polymer film laminates are now being used for electronics protective packaging, for storage, or shipment. Antistatic qualities on interior surfaces are required to meet electrical properties of Mil-B-81705, Type I specifications.

ESD can occur at any point in the manufacturing process. Static charges on personnel and their clothing, and on plastics in the workplace necessitate static protective materials to lower device failure. Statically conductive thermoplastics were introduced in the late 1970s and were designed for dissipating electrostatic charges on molded parts. The compounds have a surface resistivity ranging from 10^4 to 10^6 ohms/square. The desired resistivity has been introduced into reinforced nylon 6/6 compounds, for example, where they are expected to find increased use on microelectronic components for automobiles.[30] Federal test standard 101C, Method 4046, measures static decay rate.

There are many industrial processes where it is necessary to remove electrostatic charge accumulation from the vicinity of flammable or explosive dusts or gases. The uncontrolled discharge of static electricity is hazardous and some explosions can be attributed to this. Anti-static conveyor and transmission belts require the use of semiconductive rubberlike materials to eliminate and ground electrostatic charges. Although other means such as contact points, radioactive static eliminators, and water sprays are used, the anti-static belt is the most satisfactory in action.

Hoses carrying flammable powders and liquids should be prepared from semiconductors for safety reasons, and rubberlike materials are employed. In hospital operating rooms, the hazards accompanying most anesthetics can be reduced by semiconductive equipment such as hoses and semiconductive floors. In the latter example, metallic-filled and carbon black flooring compounds have been successfully used to bleed off static electricity. In fact, the processing of photographic film and the assembly of precision electronic devices represent work areas in which there are accumulations of dust and lint which would be detrimental to the end products. Semiconductive walls and flooring have been successfully prepared from metallic and graphite-filled plastic compositions. These surfaces prevent electrostatic charges from accumulating and from attracting dust and lint particles, thereby enabling air conditioning systems to remove the contaminants.[31]

Recent examples of anti-static films composed of metal/polymer composites possess an intermediate surface resistivity, which is greater than the lower resistivity required for EMI/RFI shielding. However, the 10^6 ohms per centimeter volume resistivity is 10-orders of resistivity lower than required, when polymer materials are selected for high resistivities that characterize good electrical insulation. Some metal/polymer composite films have been used in the packaging of food products because of their exclusion of moisture and oxygen.

Miscellaneous recent developments related to anti-static films appear in the following paragraphs:

Antistatic films have been prepared by introducing electrically conductive fine particles of metallic oxides of zinc, Sn, Si, In (particle sizes 0.01 to 0.7 μm) into oligomers containing polyvinyl and polyvinylidene groups. The specific volume resistance was 10^5 ohm-cm or less.[32]

EMI shielding was obtained from films of polyvinyl chloride and polyvinylidene chloride by treating the film in a liquid amine to dehalogenate the PVC and lower resistivity. At least one-half mole of a liquid amine per mole of halogen moiety was used, at temperatures of up to 200°C. Among the amines used were diethylene triamine and triethylamine. The transparent film became yellow, light brown, and finally a deep black irridescent color. When laminated to a cast epoxy resin, a semiconductive surface was obtained with a specific volume resistance of 10^8 ohm-cm of 25 μm PVC film. The bleed-off of static electrical charges from organic films was much more rapid than in the presence of an untreated film.[33]

Antistatic composite films are good for packaging electronic components, pharmaceuticals, garments, and for the packaging of foods. The high resistance of aluminum foils to penetration by oxygen and to water vapor makes such materials likely candidates for the preservation of foods. There are anti-static organic complexes which are effective in lieu of thin metal foils. Thus, for example, polyolefin films such as high-density polyethylene are blended with less than one percent hydroxyl-2-alkyl imidazoline and molded at 200°C into a 20 μm film. When coated with fluoroalkyl carboxylate at 0.2 grams per square meter, a surface resistivity of 10^{10} ohms per square is obtained.[34]

Aluminum vapor deposited on unstretched polypropylene film (30 μm) at a thickness of 1 to 20 nm exhibits good resistance to oxygen and to moisture penetration, even after stretching. An amorphous layer of aluminum oxide is formed on the aluminized surface, which retains its anti-static qualities.[35] For packaging static-sensitive materials, laminates of 1–7 mm foam sheets containing anti-static agents are prepared with polyethylene terephthalate films which have been coated with metal (such as aluminum).[36]

The Nuclear Regulatory Commission has issued an order curtailing the use of certain static eliminator devices that use microencapsulated polonium 210 coated with nickel. Called ionizing air guns, these devices have been used to suppress static electricity in industrial processes. Some of the radioactive polonium capsules have been contaminating surrounding areas.[37]

Conformal polymer coatings on high voltage equipment which operate in a high vacuum environment have specific requirements for a semiconductive coating which will bleed off unwanted positive or negative charges accumulating on the surfaces of electrical insulators. High voltage components may be bombarded with electrons originating from the high voltage field itself, from nuclear particles, or from outside sources in space such as high energy gamma rays and cosmic rays. Ionization paths may form on the insulator surfaces, resulting in flash-overs between neighboring conductors. The maximum surface resistivity of 10^{12} ohms per square is derived from an epoxy/polyamide coating system (less than 0.25 mm thick) with boron powder having a particle size of less than 20 μm, and approximately 50% by weight. Boron powders are suitable in a nuclear environment.[50]

There are chemical additives which provide good anti-static properties to films without metal additives. Acrylic resin compositions with permanent anti-static properties have been prepared by copolymerization of methyl methacrylate with a vinyl monomer with the inclusion of a sulfonate, and a specific alkylene oxide and/or acidic phosphate. Cast transparent films with surface resistivities of 10^9 to 10^{10} ohms per square are reported with tensile strengths of 600 to 700 kg/cm^2.[51]

PLASTICS PACKAGING FOR FOOD PRODUCTS

Polymer films play important roles in the packaging of food products. To achieve their success the co-extrusion of more than one film—up to nine layers of resins and adhesives—have been used. The presence of multi-polymers in these composites is necessitated by the fact that some polymers have a poor performance in a few environments, while they do well under other conditons. Resistance to oxygen and carbon dioxide is difficult for films such as low density polyethylenes (LDPE), while such films perform well along with polypropylene and HDPE as moisture barriers. Characteristics of selected barrier films are noted in the accompanying table. Developments in aluminized polymer films are extending the fields of applications of polymers to food packaging because of the combined gas and moisture resistance of their aluminum films. These techniques were orginally designed for electronics applications and as static eliminators in packaging electronic components. Hence, in this section, attention is directed to metal/polymer composites for specific roles in food packaging. The biggest deterrent to the use of these composite films is the lack of transparency of the aluminized polymer films, preventing visual inspection of the contents of the package.

A brief outline of the broad spectrum of polymer film performance appears in Table 7-6, abstracted from Schreiber's article.[38] Aluminum film is included for comparison.

It would appear that composites of aluminum films and LDPE have advantages in everything but transparency and crease resistance. New developments in polymer films continue to appear and will affect consumer acceptance—packaging is a large market, over $8 billion in the United States—and the likelihood of metal/polymer composites assuming a respectable position is good.

Table 7-6 Characteristics of Barrier Films Resistance to Vapors[38]

POLYMER FILM	1985 PERCENT OF FILM MARKET	TEAR STRENGTH	OXYGEN	MOISTURE	OIL AND GREASE
LDPE	37	High	Poor	Good	Fair
HDPE	27	High	Poor	Good	Good
Polypropylene	7	Low	Poor	Good	Good
Polyethylene Terephthalate	5	Moderate	Good	Fair	Good
Nylon	1	Moderate	Good	Poor	Good
P.V. dichloride	1	Low	Good	Fair	Good
Aluminum film		Low	Excellent	Excellent	Excellent

One of the principle deterrents to aluminum foils in food packaging was that the resistance offered to gases fell appreciably during multiple creasing. Tough polyester films or LDPE films, and thin aluminum films were necessary for good barrier resistance to oxygen. Metallized (Al) biaxially oriented polypropylene films have, during the past year, gained a significant part of the fast food market (candy, cookies, granola, etc.).[39] Ten to twenty micrometer films of aluminum are being evaluated in polymer film composites for packages requiring microwave cooking—focusing heat on the contents which require browning or crisping.

Metallized films in the packaging industry have been reviewed by Skodis.[52] Barrier films must hold up under severe creasing, and under these conditions aluminum foils are not satisfactory. Vacuum deposited aluminum on polyester film has a good performance record. Metallized paper has been successful on labels for beverages and paint supplies through control of moisture content in the paper base. It was noted that metallized biaxially oriented polypropylene films have a good share of the flexible snack food market (candy, granola, cookies, etc.). Non-food packaging includes multi-wall bags for shipping, chemicals, and packages for photo film.

High barrier metallized laminates for food packaging were prepared from two metallized films by Camvac (U.K.). Barriers were developed to water vapor and to oxygen, with superior fold resistance. Oxygen permeability was .02 mg/m^2/day and moisture permeability was .05 g/m^2/day.[54]

The pouches have been laminated from spin-bonded polyethylene, a layer of aluminum foil for shielding, and an anti-static polyolefin film. A serious reliability problem, which has faced the electronics industry, has been electrostatic discharge (ESD) damage to sophisticated microelectronic components and assemblies. Some of the modules which need the protection from ESD have cost $20,000 to $30,000. These critical electronic assemblies are programmed for use on submarines, satellites, and aircraft.

Metallized papers and foils are being offered for the protection of electronic components. Several major material suppliers have offered these pouches at recent trade shows. Shielding containers have also been prepared from a chlorinated butyl rubber containing 20 to 75% powdered lead flakes for protection of electronic devices from sound, radiation, and vibration. The lead powder passed through 100 to 325 mesh.[53]

For decorative gift packaging, photofilm packages, and some shipping bags, polyvinyl dichloride has been aluminized, thus offering metal/polymer composites in this area. It is necessary to avoid overdrying and adjust moisture in the cellulosic paper layer, to avoid curling in multiple composites of aluminum, paper, and polymer.

COAXIAL CABLE SHIELDING

Coaxial cables for electrical transmission consist of an inner central conductor surrounded by a tubular outer conductor. The layers are separated by polyethylene spacer disks which maintain the symmetry of contruction. At high frequencies, signal currents concentrate near the inside surface of the outer tubular conductor. Decreasing the thickness of this sheath improves the cable's shielding qualities and increases transmission loss. Conductors are usually copper or aluminum, with low loss polyethylene as the dielectric.

Metal/polymer composites are used in coaxial construction, particularly in flexible coaxials which are used extensively in communication systems. Multiple constructions provide 132,000 telephone channel capacity over 10 working coaxial pairs (Fig. 7-7). Additional protection is provided by impregnation of the sheath with thermoplastic cement in the construction of multiple coaxial cables. Coaxial systems along with microwave radio, provide most of the long distance communications in the United States.[55]

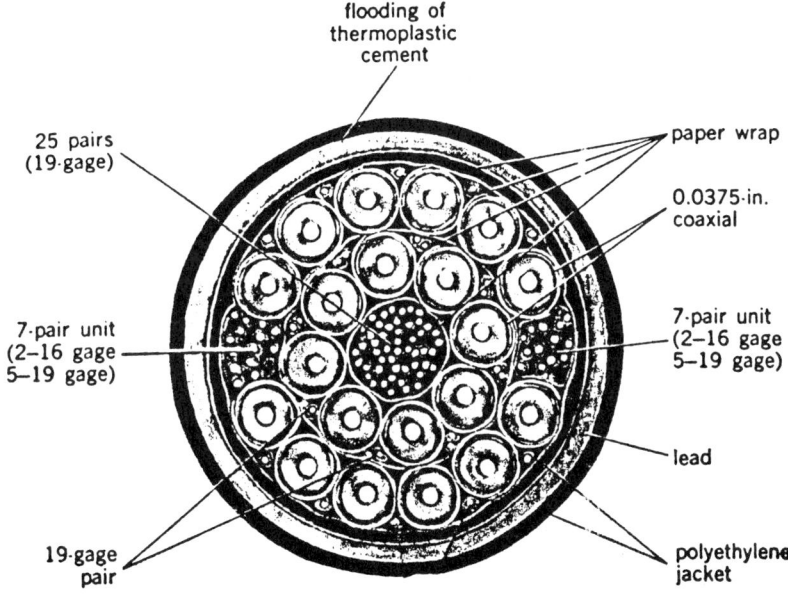

Figure 7-7. Multicoaxial transmission line.[55] Courtesy of McGraw-Hill Publ. Co.[55]

SHIELDING AGAINST X-RAYS AND NUCLEAR REACTORS

The character of shielding for x-rays and nuclear reactions is distinct from the shielding discussed in this chapter primarily because thick layers of heavy metals may be involved. Lead blocks are effective shields against x-rays, though composites of powdered lead are generally dispersed in an elastomeric polymer to prepare clothing and flexible sheets. These accessories are found in most x-ray laboratories. Gamma radiation and low energy cosmic rays may also be reduced by large masses of the heavy metals to protect personnel and equipment.

Nuclear reactions in a nuclear power plant release large quantities of neutrons covering a broad spectrum of energy. While graphite moderators are designed to absorb much of this energy, composite shields have been devised to shield instruments and personnel from the deeply penetrating neutrons.[40]

For gamma radiation shields Table 7-7 indicates the thickness of metal required to reduce radiation by one-half for gamma energy in the 0.1 to 5.0 M.E.V. range.[41] The developments of metal/polymer gamma shielding may be guided by this data.

Table 7-7 Thickness of Material (in cm) to Reduce Gamma Radiation by One Half[41]

MATERIAL	GAMMA ENERGY (M.E.V.)					
	0.1	0.2	0.5	1.0	2.0	5.0
Aluminum	1.60	2.14	3.50	4.17	5.92	9.11
Iron	0.26	0.64	1.07	1.49	2.09	2.84
Copper	0.18	0.53	0.95	1.33	1.86	2.47
Lead	0.012	0.068	0.42	0.90	1.34	1.44
Water	4.14	5.10	7.17	9.82	14.05	23.02
Air	35.5	43.6	61.9	84.5	120.5	195.8
Concrete	1.75	2.38	3.40	4.65	6.60	10.28

REFERENCES

1. Bradish, F., *Conductive Composites for Shielding*, SPI Composites Institute, 33d Annual Conference, Section 4A, (1978).
2. Simon, R., *Shielding Through Conductive Plastics*, S.P.E., New York: Antec, (1980).
3. Delmonte, J., "Minimizing Radio Disturbances on Aircraft," *Aero Digest*, (Jan. 1935).
4. Roberto, O. and Davidson, T., *EMI/RFI Shielding Plastics* S.P.E. Retec Meeting Proc., p. 104, (June 1984).
5. *Electromagnetic Shielding Effectiveness in Planar Materials*, ASTM, ES7, (1983).
6. Bigg, D., "Electrical Properties of Metal Filled Polymer Composites," In *Metal-Filled Polymers*, Bhattacharya, S.K. (Ed.), New York: M. Dekker (1986).

7. Stephens, K., *Plastics Engineering*, 43: 33, (Feb. 1987).
8. Thorpe, M., *Mod. Plastics Encyclopedia*, p. 384, (1988).
9. Anon., *New Scientist*, p. 24, (Aug. 30, 1984).
10. Gerteisen, S., *Mod. Plastics Encyclopedia*, p. 379, (1988).
11. *Eccoshield Bulletins on Shielding and Conductive Systems for Electronics*, 1988 Emerson-Cuming (Grace Co.), (1988).
12. Anon., *Plastics Compounding*, p. 19, (Jan./Feb. 1980).
13. Liu, N., and VanderMeer, R. to General Electric Co., U.S. 4,566,990, (Jan. 28, 1986).
14. Liu, N. to General Electric Company, U.S. 4,596,670, (June 24, 1986).
15. McLean, M., D. Sharp, E. Hartman (Allied-Signal) SAMPE, 33d International Symposium, p. 634, (March, 1988).
16. Cordon, J. and Bitzer, T., *32d SAMPE International Conference*, p. 68, (April 1987).
17. C.A. 100-157901a, Toshiba Chemical, J.P. 58,220,714, (Dec. 22, 1983).
18. C.A. 100-140892a, Takeda Chemical, J.P. 58, 145,769, (Aug. 30, 1983).
19. C.A. 101-132,008g, Kuraray Co., J.P. 59,86,638, (May 18, 1984).
20. C.A. 101-132,108c, Allied Corp., U.S. Applic., 436,238, (Oct. 25, 1982).
21. C.A. 104(16)131551u, Asahiper Co., J.P. 60,177,080, (Sept. 11, 1985).
22. Smoluk, G., *Modern Plastics*, P. 48, (Sept. 1982).
23. C.A. 105-134999c, Mason, J. and Zaka, Y., *Trans. Inst. Metal Finish*, 64(3): 110, (1986).
24. C.A.105-154370u, Topan Print Co., J.P. 61,127, 197, (June 14, 1986).
25. C.A. 106(18) 139504I, Nippon Zeon Co., J.P. 61,249,713, (Nov. 6, 1986).
26. Stephens, K., *Plastics Engineering*, 43:33, (1987).
27. C.A. 108(6) 39311a, Showa Rubber Co., J.P. 62,212, 465, (Sept. 8, (1987).
28. Asahipen, J.P. 60,177,080, (Sept. 11, 1985).
29. C.A. 106(16)134,899v, Mitsubishi Rayon, J.P. 61,66-755, (April 5, 1986).
30. Nangrahi, K., and Geiteisen, S., *S.P.E. Retec on Shielding*, p. 54, (June 1984).
31. Delmonte, J., *Metal Filled Plastics*, p. 84, New York: Reinhold Publishing Co., (1961).
32. H. Kawaguchi, Fuji Photo Film Co., U.S. 4,571, 361, (Feb. 18, 1986).
33. Gurgio, A., et al. to Dow Chemical Co., U.S. 4,643, 953, (Feb. 17, 1987).
34. C.A. 108(6)39288y, Showa Denko, J.P. 62,141,040, (June 24, 1987).
35. C.A. 108(6)39259q, Toyo Metallizing, J.P. 62,179, 935, (August 7, 1987).
36. Raszewski, L. to Crowell Corp., U.S. 4,698, 254, (October 6, 1987).
37. Anon., *Chemical Engineering News*, p. 15, (February 22, 1988).
38. Schreiber, P., "*Plastics Packaging,*" Plastics Engineering, 42: 29, (May 1986).
39. Skodis, L., *Packaging Encyclopedia*, p. 35, (1988).
40. Delmonte, J., *Radiation Resistant Plastics* Chicago: S.P.I. Reinforced Plastics, (Feb. 2, 1965) and Peters, J., U.S. 3,114,839, (Dec. 17, 1963).
41. Henry, H., *Radiation Protection*, p. 485, New York: John Wiley, (1969).
42. Lesurf, J. and Sutton, C., *New Scientist*, p. 55, (April 21, 1988).
43. Gerteisen, S. and Nangrani, K., *Wilson-Fiberfil Tech. Report on EMI/RFI Shielding*, (1987).
44. Beacham, H. and Muconicks, P., to FMC Corporation U.S. 4,552,687, (Nov. 12, 1985).
45. C.A. 109(37)23870s, Kawatetsu Techno Research, (July 25, 1988), J.P. 63,61,053, (March 17, 1983).
46. C.A. 107(16)135339r, Mitsubishi Monsanto Chem. Co., J.P. 62,119,249, (May 30, 1987).
47. C.A. 107-135 5531V, Idemitsu Petro Chemical Co., and J.P. 62,101,654, (May 12, 1987).
48. International Paper Co., *Techn. Bulletin on C-Fiber Mats*, (1985).
49. Liao, S.,*Microwave Devices*, Englewood Cliffs, NJ: Prentice Hall, (1980).
50. Jolitz, W. and Williams R., (Ford Aerospace), U.S. 4,639,385, (Jan. 27, 1987).
51. Kato, Y., et al (Sumitomo Chemical), U.S. 4,699,964, (Oct. 13, 1987).

52. Skodis, L., *Packaging Encyclopedia*, p. 35, (1988).
53. C.A. 108(6)39311a, Showa Rubber Co., (Feb. 8, 1988)., J.P. 62,212,465, (Sept. 18, 1988).
54. C.A. 107(38)107405J, Kelly, R., *Jl of Plastic Film Sheeting*, 3(1): 41, (1987).
55. Anon., "Coaxial Cable," *Encyclopedia of Science and Technology,* 4:98, New York: McGraw-Hill, (1987).

8

METAL/POLYMER COMPOSITES IN MAGNETIC COMPONENTS

Metal/polymer composites are evident in sub-micron electronic components. Though they do not have the structural visibility of composites of metals and polymers described in the two previous chapters, they are major factors in the economy of the United States. For example, virtually all magnetic recording media now in use consists of magnetizable particles dispersed in an organic polymer. The total United States market for magnetic particles in 1985 was approximately $285 million, and the market for tapes and disks was $8 billion. There are many developments taking place for which theoretical analyses are incomplete, as will be observed in this chapter. The physical and electrical properties of materials developed at a macro-level are not always applicable at sub-micron dimensions. The interactions of energy levels of unpaired electrons, the influence of the broad spectrum of radiation, and hostile environments become the focus of attention at sub-micron distances. Semiconductors, magnetic devices, storage batteries, and capacitors are, in many instances, examples of micro-metal/polymer composites. Magnetic recording devices and magnetic components comprise important segments of the electronic industry and will be examined in this chapter. The roles of polymer enamels used with magnetic wires will be summarized. Chapter 9 will concentrate upon the semiconductor devices and energy storage accessories.

Among the more exciting applications of advanced magnetic devices are the experimental trains being developed in Japan and in West Germany. The new train is called a MAGLEV, a contraction of magnetic levitation at high speeds. The trains literally float on air, unhindered by friction except for wind resistance. They are supported by and propelled by the force of powerful magnets. Alternating currents in the stationary magnets of the guideways change the polarity of and interact with the magnets on the train. Speeds of over 300 miles per hour have been attained on a straight track in Japan, and the German model (the TR-07) is designed to reach this speed (Fig. 8–1).

There are basic differences between the MAGLEV designs of West Germany and Japan. The Japanese model, which is based on superconducting electromagnets installed on the train, will operate at very low temperatures. Super-

188 METAL/POLYMER COMPOSITES

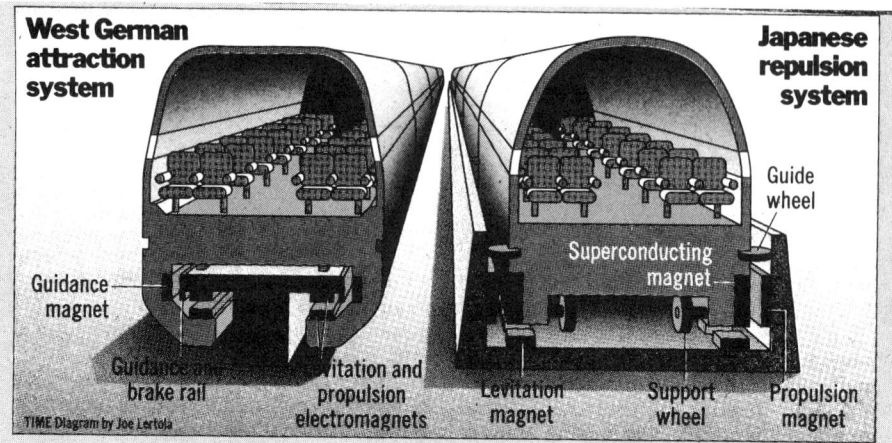

Figure 8-1. Experimental high speed trains designed with magnetic levitation. Courtesy of *Time Magazine*.[1]

conductors will cut energy losses and use liquid nitrogen as a coolant, instead of liquid helium. The West German design uses conventional electromagnets activated by a distant control station, to establish a polarity opposite to the magnets on the train. The attraction lifts the train about $3/8$ inch, precluding the conventional rails used for present trains. At present, no corresponding developments are taking place in the United States, though a major project is underway that will use the West German train on a Los Angeles to Las Vegas link.[1]

These are some of the current developments, and in this chapter, the metal/polymer composites will be identified. Comparative tables of selected soft magnetic materials and of representative hard permanent magnetic materials will be shown. The roles of polymers with magnetic materials are evident by their presence in tape recordings. Hence an overview of materials in communication devices will be presented after the basics of magnetic materials are reviewed. Prospects for polymer/metal magnetic composites are enhanced by the following additional factors:

- The exciting properties of new magnetic materials prompts interest in potential applications. Many of the alloys, such as "Alnico" are limited by their brittleness, hardness, and difficulties in machining. The reduction of these hard materials to a powder form, their molding and shaping with the aid of minor additions of powdered organic polymers, followed by controlled sintering at high temperatures make possible useful fabricated components. The example of an injection molded steel gear, in Chapter 3, should be re-examined.
- Rapid solidification processing (RSP), particularly of nonconventional steel alloys, have yielded a plethora of new compounds. The RSP processing frequently leads to an amorphous, noncrystalline metallic powder. Early reports indicate properties of magnetic components prepared from some of these powders may revolutionize transformer and motor design. Metal in powder form, provides new opportunities for synergistic combinations of metals with organic polymers.
- Though not a very recent development, the effective utilization of barium and strontium ferrites are of major consequence in magnetic devices. Their presence in molding compounds is also discussed in this chapter as they fulfill many important functions in magnetic devices.

MAGNETIC MATERIALS

Plastics and metallic alloys have contributed, in no small measure, to the rapid and versatile growth patterns of magnetic materials. Both magnetically soft materials as well as hard permanent magnetic materials fulfill many functions as components for communications apparatus. To understand the significant material developments, it is necessary to examine some of the fundamentals of magnetic measurements. Magnets fit into three broad categories:

1. *Soft magnetic materials.* The magnetic field is produced by an electric current, and the materials are characterized by their low loss and their high permeability. (Fig. 8-2.) There are varieties of materials with combinations of magnetic properties, mechanical properties, and cost. Examples appear in Table 8-1. The strength of the magnetic field may be regulated by the electric current within the saturation magnetization limit of the material. Motors, transformers, and generators—requiring many field reversals per second—utilize soft magnetic materials which are rapidly magnetized and demagnetized with minimum energy loss.

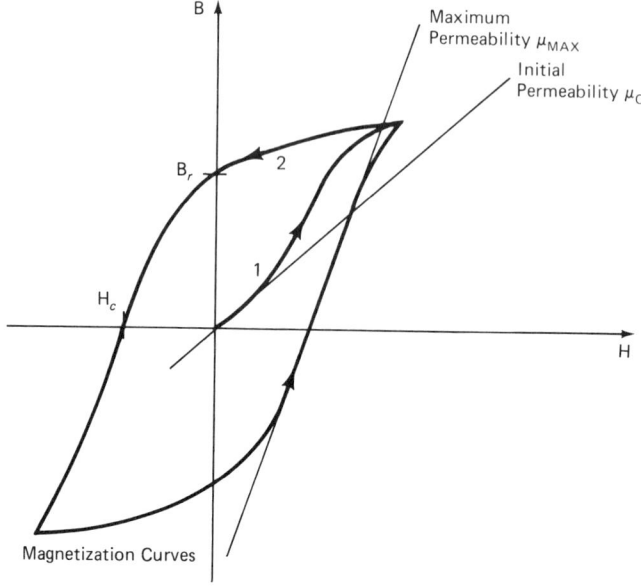

Figure 8-2. Properties related to the magnetic hysteresis loop. B_r—permanent induction after removal of the applied field H; H_c—the coercive force or reverse field required to bring induction B back to zero; μ_{max}—relative permeability, the slope of the B/H curve.

2. *Permanent magnets.* Recent developments in powdered and sintered metals have led to good candidates for hard permanent magnetic materials which resist demagnetization. Hardened steel alloys are magnetized by being placed in a strong magnetic field. They are characterized by the residual or remanent magnetism (B_r) and high coercive force (H_c). The remanence is the structural flux density of a permanent magnet, described in units of gausses (one gauss $\times\ 10^4$ equals one Tesla). The coercive force, defined in units of oersteds, is the reversed magnetizing force required to reduce the remanence to zero and thus demagnetize the specimen. Permanent magnets are used for loudspeakers, meters, holding devices, and telephone equipment, in which approximately constant field strengths are required.

3. *Magnetic ferrites.* Magnets are also derived from gamma-Fe_2O_3 and Fe_3O_4 powder having certain grain sizes and shapes designed primarily for use in making magnetic recording media. They have good chemical resistance and low density. Barium oxide and cobalt oxide may also be combined with the iron oxides. A compound of barium and iron oxide ($BaFe_{12}O_{19}$) is marketed under the trade names of "Arnox" and "Ferroxdure." When produced at high

Table 8-1 Properties of Selected Magnetic Materials

Soft Magnetic Materials

COMPOSITION OR TRADENAME	COERCIVE FORCE H_c (OERSTEDS)	RESIDUAL INDUCTION B_r (GAUSSES)	RELATIVE PERMEABILITY (U_{max})	resistivity ohm/ CM × 10^{-6}
High purity Fe	1.10	13,000	5,000	10.
High purity Fe (annealed in H_2)	.05	13,000	200	10.
Fe, 2 silicon	0.5	10,000	7,500	41.
Permalloy 78 Ni/22 Fe	.05	7,000	100,000	16.
Mumetal—77 Ni, 5 Cu 2 Cr, balance Fe	.05	6,500	100,000	
50% Co-Fe, 2V	2.0	24,000	5,000	7.
Cobalt, pure	10.00	5,000	250	
Nickel, pure	0.7	4,000	600	6.8
Amorphous Fe-B-Si (Metglas 2605 S-2)	.03	(15,000)	300,000	130.

Permanent Magnetic Materials

COMPOSITION OR TRADENAME	COERCIVE FORCE H_c (OERSTEDS)	RESIDUAL INDUCTION B_r (GAUSSES)	MGOE	MAX. ENERGY (BH) MAX. kJ/M^3
Steel (0.9% C)	70	10,300		
Steel (3.6% C)	240	9,500		
Alnico 5—50 Fe, 24 Co, 15 Ni, 8 Al, 3 Cu	620	12,800	5.2	42
66 Fe, 33 Nd, 1B, ($Fe_{14}Nd_2B$)	11,000	12,300	36.0	290
$SmCo_5$ (60-Co-34Sm)	640	8,700	18.0	144
11.5 Co-33Cr-Fe	60	13,000	5.2	42
Barium Ferrite Ba 0-6 Fe_2O_3	2,100	4,300	4.5	36

Sources of Data: McGraw Hill Encyclopedia of Science and Technology, Vol. 10, p. 294—1987.[2]
Condon and Odishaw—*Handbook of Physics,* Sec. 4-131, 1980.
C. Y. Chin—*Science,* 208:888, May 23, 1980.[3]

temperature under a strong magnetic field, an oriented structure with an isotropic structure forms a permanent magnet with high energy product (BHmax) of 3.2 gauss-oersted $\times$ 10^6 (25 kg/m³).[2] Both hard and soft ferrites are available. Particle size of $BaFe_{12}O_{19}$ is typically 1.7 μm.

Magnets are characterized by parameters that measure the strength and permanence of their magnetic fields. A material is magnetized in a magnetic field generated by an electric current flowing through a coil. Magnetic field intensity or magnetic force (H) is measured by the force acting on a unit pole. Unit field intensity, the Oersted, is that field which exerts a force of one dyne on a unit magnetic pole. The magnetic lines of force, the magnetic induction, B, are measured in gausses. The magnitude of the reverse current required to demagnetize the magnetic lines of force is an indication of the permanence of the magnet.

Table 8-1 shows typical soft magnetic materials and permanent magnetic materials. The increase of coercive forces in new alloys during the past 10 years has influenced the expanding interest in these materials. The soft magnetic materials, particularly those with high permeability, are influencing motor, transformer, and generator designs. These devices may be built in smaller sizes for equivalent energy output. Above frequencies of about 10k Hz, some of the soft magnetic materials of FeNi Permalloy are degraded by eddy currents which generate excess heat losses in the material. Preferred materials in the high frequency range are soft ferrites with higher orders of electrical resistivity.

Because magnetic materials may be exposed to high temperatures during operation or during processing, as they are fabricated into an electro-mechanical device, information on the Curie temperature is essential. This is the temperature above which the magnetic properties completely disappear. On returning below the Curie temperature, prevailing magnetic fields (including the earth's magnetic field) would be registered in the material. Curie temperatures of selected magnetic materials are shown in Table 8-2.

The functional contributions of plastic materials to magnets may be listed as follows:

- The fabrication of brittle magnetic alloys is simplified when the particles are dispersed in a plastic matrix and pressed to shape.
- Variable magnetic characteristics become design aids since it is possible to prepare nonhomogeneous magnets with selected material distribution.
- The suspension of magnetic particles in liquid resinous media has been adopted in electromechanical clutches.
- Because of the the physical presence of plastic insulative coatings on magnetic particles the intrinsic length of the magnetic path is reduced and eddy currents are minimized.

Table 8-2 Curie Temperatures for Typical Ferro-Magnetic Materials

SUBSTANCE	CURIE TEMPERATURE (°C)
Fe	670
78 Permalloy	600
50% Co-Fe	980
Co-	1,021
Ni-	260
Fe_3O_4	485
Ba Ferrite	450
Sr Ferrite	460
Alnico 5	900
Fe-Nd-B	300

- Corrosive effects are minimized by the presence of plastic materials as coatings on the magnetic particles.
- Eddy currents and hysteresis losses may be controlled for maximum operating efficiency.
- Assembly features may be accommodated by the inclusion of bushings, inserts, or other components when the powders are compacted together. The stiffness of ceramics, fibers, and non-ferrous metals will contribute to better designs.

Markets for metal/polymer composite magnets have been growing actively outside the electronics industry. The products these markets produce include flexible permanent magnets for sealing strips inside refrigerator door gaskets; magnetic weather stripping for insulating exterior doors; advertising signs and novelties; and toys. Matrix resins include nylons, flexible polymers, polyurethane, chlorinated polyethylene, and ethylene-acrylate copolymers. Energy products for the polymer bonded magnets are in the range of 3 to 10 kJ/m^3 (BH_{max}) as compared with rigid commercial $SmCo_5$ samples which measure 130 to 160 kJ/m^3. Studies of the mechanism of coercivity in $SmCo_5$ suggest domain nucleation at the grain boundaries. The original compound, $SmCo_5$, was discovered in the mid-1960s by Karl Strnat at the United States Air Force Lab in Dayton. The high price has limited its use in large volume.

DEVELOPMENTS IN PERMANENT MAGNETS

New improved permanent magnetic materials are characterized by high levels of saturation magnetization, high Curie temperatures and strong uniaxial magneto-crystalline anistropic fields.[4] This review reports upon:

RETM$_5$ and RETM$_{17}$
RE = rare earth elements
TM = transition metal elements

Saturation magnetization levels determine the maximum level of flux obtainable from the metal, and Curie temperature must be substantially higher than use temperatures. In applications to gyroscopes, accelerometers, millimeter and microwave devices variations in flux level with temperature must be held to a minimum. High coercive H$_c$ levels are a measure of a magnet's resistance to demagnetization and such levels are necessary when the material is intended for use in environments where it will be subject to large self generated fields or to externally imposed demagnetizing fields. The report compares the performance of RETM$_5$ (SmCo$_5$) and the Fe-Nd-B magnet materials, and points out latter alloys which have lower cost and better machinability. However, these latter alloys have the drawback of a high reversible temperature coefficient of magnetization associated with a low Curie temperature of 300°C. Thermal prestabilization treatments of the magnetic materials are also discussed in this reference, as are the square hystersis loops of Sm-Fe-Co-Cu and Sm-Co-Cu alloys.[4]

New rare earth permanent magnets with maximum energy products up to 250 kJ/m³ and chromium-cobalt-iron alloys which duplicate the performance of "Alnico" 5 (8Al, 15Ni, 24Co, 3Cu and a balance of iron) are discussed in Reference 3.

LNP Corporation reported on the properties of oriented versus nonoriented nylon 6 molded magnetics (Ref. 5). Data in Table 8-3 indicates the advantages of orientation of the particles. Low viscosity resin matrices permit better magnetic orientation during molding. The maximum energy product (BH$_{max}$) is considerably lower in products from molded plastic compounds. The fabrication ease of molded magnetics must be weighed against the intrinsic superior magnetic qualities of a solid permanent magnetic alloy.

Table 8-3 Properties of Nylon 6 Molded Magnets[5]

	COERCIVE FORCE H$_c$ OERSTEDS	RESIDUAL INDUCTION B$_r$-GAUSS	MAXIMUM ENERGY BH$_{max}$ (KJ/M³)
Non-oriented nylon 6 Molded magnets	1060	1300	0.3
	1270	1560	0.6
Oriented nylon 6 Molded magnets	1630	1840	0.6
	1950	2300	1.20

The application of a magnetic field during the cure or setting of a resin binder to provide alignment of magnetic particles has been used on numerous occasions. For an example see Reference 41. In general, if the resin content is high enough to permit the physical alignment, the concentration of the resin is such as to markedly reduce the strength of the magnet. To permit greater concentrations of magnetic particles, they may be placed in a magnetic field and vibrated to assist alignment.

Ground "Alnico" alloys, as well as iron and barium oxides, are treated with resin binders and molded under heat and pressure. Good stability was demonstrated over a wide temperature range. Lower eddy currents, and more accurate dimensions for the molded permanent magnets, as well as the ability to produce more complex shapes are claimed. Compacted magnets are reported as demonstrating magnetic strengths 80 to 90% of the cast metal with good economy in the complex shapes.

A recent paper discusses problems of achieving alignment of magnetic particles in a strong magnetic field. Energy relations relating to the torques to produce orientation of nickel coated graphite fibers (3 mm length) and low volume loadings (under 1.0%) were used.[6] Higher volume loadings inhibited the orientation of the fibers because of higher viscosities.

For maximum resistance to demagnetization, the expensive non-ferrous alloy "Silmanal" is unique. Used for applications in which magnets are subject to extreme demagnetizing influences, it is usually pressed into thin discs and magnetized across the thickness. Other alloys such as the "Amico" and "Cunife" rods are fabricated into useful magnetic forms and extensively employed in instrumentation. It is most interesting to note the new horizons and directions which have been made possible by sintering techniques and by combinations with insulating binders. The net results are caused by the uniqueness of the metal alloys and oxides involved. One cannot help but be impressed by the parallel fabricating problems of metal powders and plastics powders, which may lead to integrated designs embodying insulating layers of plastics and electrically conductive or magnetized layers of metallic alloys. It is not inconceivable that products in the future will manifest coordinated electrical and insulating elements, both of which may be formed by similar fabricating procedures.[7]

Composites of polymers and magnetic particles have specific applications in electrical and electronic devices. There are many examples of new high energy magnets and energy efficient magnetic materials for motors, generators, and transformers, though the use of polymer additives for these major applications is limited. The metal/polymer composite concepts are more evident in powder metal composites for electronic devices, semiconductors, magnetic recording tapes, communication equipment, and sensitive instrumentation.

MAGNETIC MATERIAL DEVELOPMENT FOR TRANSFORMERS, MOTORS AND GENERATORS

Powerful new magnetic materials based on a neodymium/iron boron alloy were reported in 1984. The compound was first announced in 1979 in Soviet journals by crystallographers at the Lvov Ivan Franko State University in the Ukraine. Research groups in the United States and Japan have been concentrating on $Fe_{14}Nd_2B$ alloys, not only because of their good magnetic properties, but also because of the availability of lower cost raw materials, and the substantial quantity of iron used.[8] It is reported that the presence of boron enhances the likelihood of rapidly cooled materials being amorphous, allowing greater ease in preparing fine powders. The main limitation of the new magnetic material is its low Curie temperature, 300°C, as compared to over 1000°C for some cobalt alloys. At the Curie temperature, demagnetization takes place, though as the temperature approaches the Curie temperature, the effective coercive force decreases. Hence operations should take place well below the Curie temperature, up to 100°C. In the absence of applied magnetic fields, magnetic materials will reflect and retain the earth's magnetic field as they cool below the Curie temperature (Table 8-2). Polymer/metal composites will be included in new development efforts. As indicated in Chapter 3, plastics processing methods involving small amounts of polymer binder are now being employed in the fabrication of pressed powder metal compacts.

Metal/polymer composites are present in the many thousands of transformers used for electrical power distribution, and are installed in many small electrical devices. Polymers for electric insulation, laminated silicon steel sheets for electromagnetic fields, and coils of copper wire with heat-resistant insulation have demonstrated good efficiency with low hysteresis loss. Silicon steel sheets prepared by Goss in the 1930s demonstrated a substantial decrease of hysteresis loss and a large increase of permeability in the direction of rolling.[11]

The ability to reduce energy losses significantly in transformers and motors is being developed by the use of amorphous metal powders to replace grain oriented steel sheets. It is estimated that 700,000 tons of electrical steel (Si-Fe) is produced in the United States every year.[12] Improved magnetic alloys are being manufactured by rapid solidification processes, yielding amorphous magnetic powders. As discussed in Chapter 3, the powder technologies of metals and polymers are growing closer together and the roles of metal/polymer composites are increasing.

Ultra high vacuum systems and the use of ionic beams for deposition of atoms, layer by layer, combine to make possible the creation of new materials (composites). Combinations of magnetic and nonmagnetic layers are leading to an increased understanding of the propagation of spin polarization through metals, and the effects of finite thickness on the properties of the ground state and thermodynamics of magnetic materials (Reference 13, p. 3721). Granular magnetic solids (1.5 to 15 nm) were imbedded in an insulating matrix in order to explore the influence of finite size effects.

Many fundamental research papers were presented in this symposium. These papers underscore the growing interest in small magnetic particles, their influence on the subject of magnetic materials, and their application to semiconductor technology.

MAGNETIC MATERIALS IN COMMUNICATION

Magnetic materials as used in communications cover three general areas: telephonic applications, magnetic recording, and information storage. Competitive technologies are arising from electro-optical systems, and may compete with magnetic devices.

Telephone. The use of powdered iron in magnetic cores was disclosed in a United States Patent.[14] Early developments in telephonic equipment emphasized powdered alloys of high nickel-iron content which are characterized by high permeability and low losses. These qualities should prevail in devices for telephonic and R.F. cores—low losses at low flux density.

Powders prepared from brittle castings or from the hydrogen reduction of metallic oxides have been in demand. Pressing of magnetic powders and cooling processes will induce strains and may raise the coercive force, particularly of pure powders. The commercial magnetic alloy-powder cores weigh approximately 10% of the weight of earlier iron cores and hence are more acceptable for use in telephonic equipment. Magnetic particles have been coated with copper to improve electrical characteristics. They serve to reduce eddy current losses.

Magnetic cores made from mixed oxides of iron, nickel, manganese, zinc, magnesium, and barium, which are classified as ferrites, possess the physical qualities of ceramics. Because of their high resistivity, they have very low eddy current loss, in contrast to pure metal powders which require the presence of insulating plastics. The combination of high permeability and low losses make the ferrites suitable for television transformers and other components.

The permeability (B/H) ratio of ferromagnetic materials is an important consideration as large values permit reductions in component size. It is inherently not constant. However, it is necessary to achieve some degree of constancy in filter coils used for radio and telephone circuits, and loading coils used in long distance telephone transmission lines. Uniform permeability is required for fidelity of sound reproduction. While air gaps may be used in the magnetic circuit, more satisfactory results are obtained when magnetic materials are insulated with synthetic resins. Although the permeability is appreciably lower, it is constant, and from 1 to 100 gauss is typical. Eddy currents are markedly reduced when insulated particles are employed. Alloys such as "Perminvar" and "Isopernus" have demonstrated good uniform permeability. The eddy current is that portion of a core loss due to currents circulating in a magnetic material as a result of electromotive forces induced by varying induction. The uses of thin cross sections of magnetic materials and materials with higher specific resistivity result in less eddy current loss. Also to be considered is the hysteresis loss which is proportional to the power expended in a magnetic material when the magnetic induction is cyclic.

Above 10 kHz, soft magnetic alloys, such as Fe-Ni, are degraded by eddy current losses. Soft ferrites such as $GdFe_5O_{12}$ are reported satisfactory for higher frequencies.

Magnetic Recording. Magnetic recording technology draws upon disciplines in many fields of science and engineering to resolve problems at micron and submicron levels. Audio, video, and data recording on tapes and disks are important to many contemporary activities, and the roles played by magnetic recording materials represent substantial financial investments. The two principal magnetic recording processes are depicted in Figures 8-3 and 8-4.[17] In the longitudinal process a pattern of magnetic variations is imprinted on a coating of magnetic material which has been deposited on a polyester tape (PET) or on an aluminum disk. As the medium moves past the head, a signal current is passed through the head's coil, sufficient to generate magnetic fields

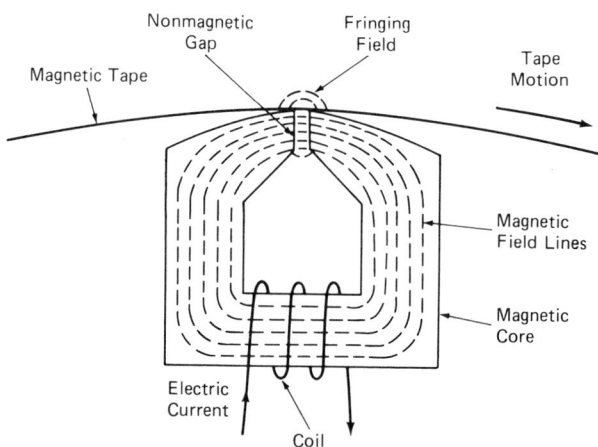

Figure 8-3. Schematic representation of a magnetic recording head. Adapted from Reference 2. Courtesy of McGraw-Hill Publ. Co.

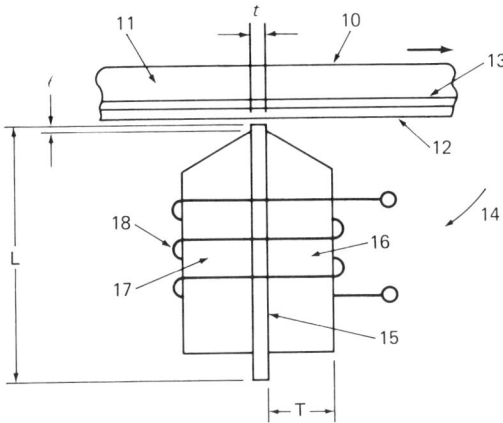

Figure 8-4. Schematic representation of Sony vertical head recording. 10—recording medium; 11—nonmagnetic material (polyimide, polyester, polycarbonate); 12—recording layer with magnetizable particles (layer 0.1 to 0.5 micrometers); 13—low coercive force layer (approximately ⅕ of no. 12—0.5 micrometers thick); 16 and 17—nickel zinc ferrite and manganese zinc ferrite poles. Courtesy of Arnoldussen (IBM) and Reference 16.

that alter the direction and magnitude of the magnetization in a way that corresponds to the information being recorded.

In digital saturation recording, the sequences are of complete magnetization reversals between positive and negative saturation values. The density of data storage depends on the sharpness of magnetization transitions with the hysteresis loop assuming more of a square pattern. The squareness ratio equals remanent magnetization (M_r) divided by the saturation magnetization (M_s). Materials for digital recording media should have squareness ratios approaching zero. In some of the notes following this section, squareness ratios are indicated.

The perpendicular recording process follows the procedures noted above with the exception that magnetization is perpendicular to the medium plane. It has been theorized that the process of perpendicular reversals permits data to be stored because of the narrower transitions on the recording medium. (Figure 8-4) Weaker signals may develop under this mode.

A magnetic medium suitable for use in a recording system employing a magnetic recording transducer or head, was developed by Sony Corporation using a perpendicular magnetizing mode.[16] The system used a polyimide film upon which was applied a low coercive force base of an alloy of Ni-Fe-Mo (18-72-7). The film was applied by sputtering at 1.5×10^{-7} torr, with an argon pressure of 2×10^{-2} torr on the base which was maintained at 250°C. The H_c of this layer is about 5 oersteds, with a saturation magnetization of 600 gauss. About 0.5 μm was formed in 20 minutes. On top of this layer was sputtered the cobalt-chromium (82-18) alloy magnetic recording layer under the same vacuum conditions. This layer was deposited to a thickness of 1.0 μm. The recording layer had a coercive force of 1500 oersteds (H_c) and a saturation magnetization of 390 gauss. The low coercive force layer was specified at no greater than 20% of the coercive force of the magnetic recording layer.[16] Figure 8-5 depicts the frequency response characteristics versus the recording density of a magnetic tape prepared by the technique outlined above. The tape displays an intermediate layer of low coercive force with a layer of Co-Cr alloy for recording. Tests were recorded at a speed of 9.5 cm/second. The single pole recording head was prepared from nickel-zinc-ferrites. Reproduction or playback of recorded signals is effected by a ring-type magnetic head with a gap length as low as 1 μm.

Tapes, rigid disks, and flexible (floppy) disks are the recording media. While

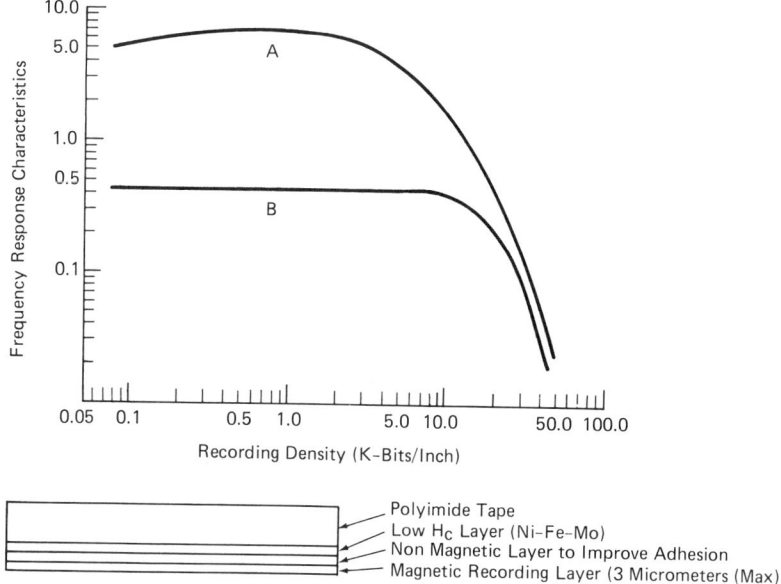

Figure 8-5. Frequency response vs. recording density. A—magnetic recording layer with low coercivity layer as intermediate; B—magnetic recording layer with no intermediate low H_c layer. Adapted from Reference 16.

historically analog signals have been recorded for audio, video, and instrumentation purposes, the trend is toward digital storage of information, as in a computer. Rigid disks of aluminum alloy are used at high speed rotation (3,600 rpm) and the recording/playback head is designed to fly above the disk surface at 0.30 μm. Flexible disks usually have magnetic coatings on both sides. They are not capable of the high density storage of rigid disks.[17]

Flexible magnetic recording media with CrO_2 particles in the magnetic material have improved considerably their mechanical stability by using a polyester-polyurethane resin binder.[19]

Current trends in magnetic storage media (e.g., tapes, disks) are toward higher packing densities, allowing large amounts of data to be stored in a small space. A polymer matrix or binder is required to maintain the magnetic particles with a given orientation, and to keep the particles adhering to the substrate.

Good adhesion and interaction between the resin binder and the magnetic particles are very important. Magnetic oxide particles are generally in excess of 70% of the coating by weight and as much as 50% by volume. The chromium dioxide (CrO_2) addition improves the frictional stability of the magnetic layer. A slurry is prepared of magnetic particles dispersed in a solution of the polyester-polyurethane binder, which is then coated on to a flexible substrate to form the magnetic recording tape.

The presence of the acicular oxide fillers contributes to higher coercivity and more efficient operation. Cobalt-doped gamma iron oxide and needle-like particles (particle size—0.4×0.07 μm—Pfizer 2,566) have been studied in the reinforcement of polyurethane (Northane CA 310) for use on magnetic storage disks and tapes. There is an interaction between the fillers increasing the tensile modulus and glass transition temperature, with a 25% volume loading, and decreasing tensile strength and elongation.[10] Figure 8-6 illustrates the relative tensile modulus enhancement (as much as 35%) at different volume fractions. Dispersing agents, such as lecithin, are also used.

For general purpose recording, gamma Fe_2O_3 with a coercivity of 300 oersteds is generally used as the magnetic medium. Particle size is about 0.4×0.07 μm. For high quality audio and video recording, magnetic particles using CrO_2 and cobalt impregnated gamma Fe_2O_3, with a coercivity of 600 oersteds have been successful. High recording density has led to the need for high coercivity recording media. Thin polymer composites are now available with 1,500 oersteds coercivity.[9]

Magnetic recording media require a thin, tightly adherent protective coating to minimize their susceptibility to wear under the record/playback transducer heads. Wear of thin magnetic films is apparent with video tape and electronic cameras. Preferred thicknesses of the protective coverings are in the 5 to 25 nm range. Perfluoropolyether polymers were developed for this purpose.[18]

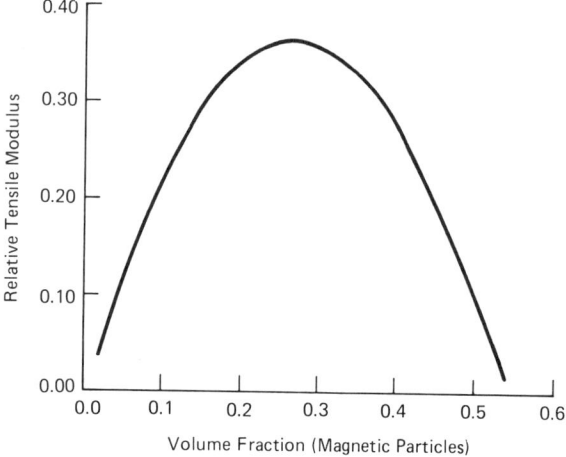

Figure 8-6. Enhancement of tensile modulus of polyurethane by magnetic fillers. Adapted from Reference 10.

Table 8-4 Characteristics of Magnetic Ferrites[20]

FERRITE CONTENT		B_r (GAUSS)	H_c (OERSTEDS)	BH_{max} (GAUSS-OERSTEDS $\times 10^6$)
VOL %	WT %			
20	40	375	400	0.08
40	76	775	730	0.10
60	88	1450	1225	0.46
75	94	2100	1375	0.94

Early efforts to increase the coercivity of gamma Fe_2O_3 finely divided particles were successful, incorporating 2 to 3% cobalt in the lattice. However, the composite demonstrated high sensitivity of coercivity to temperature changes. A solution to this problem was found by TDK Corporation (Japan) in 1974. They impregnated only the surface of the iron oxide particles with cobalt. Coercivities of 550 to 750 oersteds are now achieved in high performance audio and video tapes and in high capacity flexible disks, according to White.[15]

The cost of permanent magnets is significantly lowered by the use of gamma Fe_2O_3 and oxides of barium, lead, and strontium, thus obviating the need to use expensive elements such as samarium and cobalt. The general formula is $MFe_{12}O_{19}$, where M is a divalent metal ion or combination of metallic ions. Typical relations between the magnetic properties and the ferrite content are shown in Table 8-4 (Stackpole Corp. data).[20]

The production of magnetic fillers and methods for producing ferrite powder and flexible magnets are described in Reference 21. BH_{max} is increased when the particles are pre-oriented in their polymer binders. The ferrite-polymer mixture is pressed and then fired to bake out the binders; this is followed by sintering at higher temperatures.

Dispersions of magnetic particles in a resin which is coated on a disc or tape are the basis for magnetic recording devices. The yield point and the fatigue properties of polymer binders are important factors in the durability of magnetic coatings.[42] Magnetic tapes with an elastomeric polyurethane are influenced by acid catalyzed hydrolysis which affects the magnetic particles and the adhesion to polyethylene terephthalate tapes. For high densities of data storage CrO_2 particles demonstrate higher anistropy than needle-like gamma Fe_2O_3 particles, though the CrO_2 particles exhibit lower Curie temperatures (120° to 125°C)[42] than the iron oxides.

Dispersibility of magnetic particles was correlated with the absorption behavior of the polymer molecules. According to Nakamae, when the number of polymeric hydrophillic groups was below saturated absorbance, coagulation of the particles could occur via magnetic attraction or hydrogen bonding. When absorbance reached saturation, dispersibility of the particles improved.[43]

Magnetic recording disks generally require a thin layer of liquid lubricant—generally 3 to 5 nm—for the best protection of the slider-disk interface. A typical thin film disk consists of an aluminum substrate on which a 10 to 20 μm nickel-phosphorus sublayer and a 20 to 50 nm magnetic layer are deposited. On top of the magnetic film, a layer of sputtered carbon and a layer of liquid lubricant are deposited.[44] The physical dimensions emphasize the submicroscopic trends of metal/polymer composites in the design of advanced electronic devices.

The Division of Polymer Chemistry of the American Chemical Society has recently sponsored meetings on polymer applications in the fields of recording and data storage. Prominent in the presentations at these meetings were discussions of the physicochemical aspects of magnetic recordings. In addition to the polymer/metal composites of magnetic recordings, attention was given to the photochemical aspects of optical recording. In the latter technique, a laser beam is focused through a transparent substrate (such as polycarbonate) to effect a small temperature rise in the recording medium, producing a physical or chemical transition in the polymer. Recovery of the recorded data is accomplished by scanning the data track with a continuous low powered beam.[45] Dye-polymer mixtures, photochromic liquid crystal polymers, and photo-induced polymerization and isomerization were among the procedures evaluated. Developments of substrate materials for nonerasable and erasable optical memories have to take into account the high requirements for the level of birefringence, water absorption, stability, processability, and cost.

Data Storage. Magnetic devices lend themselves to the permanent storage of massive amounts of data because they offer much greater memory capacity at a lower cost per bit of data stored than semiconductor devices. A typical rigid-disk device stores 800,000 characters (equivalent to 400 typewritten pages) in a square centimeter of a recording medium. Some disks and most tapes can be removed and filed without destroying the data on them, subject primarily to the shelf life of the recording polymer film. As discussed earlier, needle like particles of iron oxide (gamma Fe_2O_3) exhibit the desired properties of high magnetization and high coercivity. In a polymeric binder with high viscosity, the task of dispersing particles uniformly is made easier.

Magnetic media for computer digital storage are tapes designed for archives and mass memory. Direct access storage devices (DADS) are common. Attention is called to the use of electroless Ni-P coatings (15 to 25 μm) (See Chapter 5) to increase the composite hardness on rigid Al-Mg disks.[17] In rigid disks and tapes magnetic particles in a solidified polymer base are oriented in the direction of the recording track to improve hysteresis squareness, and therefore signal amplitude and writeability.

New magnetic materials and more advanced recording heads are expected to yield a fivefold improvement in the 1990s. Lasers are being developed to read (retrieve), write (store), and erase information on magnetic films with prospects of higher data storage density.

Magnetic-optical writing takes advantage of the fact that laser heating momentarily lowers coercivity, while the reading depends on the Kerr effect, the rotation of a plane of polarized light by a magnetic field.[22] The typical magneto-optical film is a 10 nm thick layer of iron-terbium-cobalt or gadolinium-iron-cobalt. Increasing recording density, as noted by Metzer, has required a narrower gap (now at less than 1 μm). Random magnetic domains can be aligned on crystalline garnet to provide up/down magnetic bubbles. Bubble structures, depending upon a permanent magnetic field, and the lithographic techniques developed for semiconductors are viable candidates for upcoming technology in data storage.

Bubble domains are small, magnetized mobile regions within films of certain magnetic regions. The presence of a bubble can represent a binary bit of information, a one, and its absence, a zero. The bubbles may be scaled down to size, less than 0.5 μm. Magnetic garnet films have been the principle bubble material. PbO fluxed melts (molten solutions) are the media for which epitaxial garnet films are prepared.[23]

Optical data storage discs have been recently introduced into the marketplace for data storage. They are similar to the audio-compact disks used to market prerecorded music. Modulated signals are coded into the polymer substrate (usually of polycarbonate or polymethyl methacrylate). Ultimately, optical disc storage systems will be erasable and reusable. Dye-in-polymer preembossed data bits are read by a laser beam.[24] Sony's erasable optical disk drive alters the intensity of its laser beam, depending on the job to be done. To record data, the laser shines intensely to change the polarization of a magnetic field on the disk. To read the data, the laser shifts to low intensity. To erase, the laser sends out a diffused beam, eliminating the data.

MISCELLANEOUS MAGNETIC MATERIALS OF POLYMER/METAL

Polypropylene (20%) containing barium ferrite (80%) is mixed in an extruder at 110° to 150°C and melt spun through a screen to give acicular magnetic fibers.[25]

An adhesive sheet containing powdered magnetic alloys is applied to a parent metal to form magnetic alloy layers. The magnetic powders are blended with an adhesive binder in a slurry. Drying in place on a film takes place in a magnetic field.[26]

Composites of magnetic powders in polymer binders have unique characteristics which are being evaluated by performance measurements. Magnetic recording tapes are reported to have improved performance with a polycarbonate-urethane binder. A comparison table follows[27]:

CHARACTERISTIC	NEW POLYMER BINDER	EARLIER FORMULATIONS
Squareness ratio (residual to saturated magnetizations)	0.83	0.24
Luminescence signal to noise	1.5 dB	0
Chromatic signal to noise	2.0 dB	0

Mixtures of polyurethane, epoxy, and vinyl copolymers are used as binders of cobalt containing gamma iron oxide, Fe_2O_3. The magnetic material and binder were coated to 6 μm thickness on a polyester film and cured at 60°C. Good durability was reported.[28]

Magnetic coatings were prepared with ferromagnetic powders of Fe-Co-Ni (25 nm and aspect ratio of 8). Vinyl copolymer binder and polyurethane elastomers were used. A 3 μm magnetic layer was deposited on a 10 μm polyester film. This was cured at 50°C for 24 hours. The squareness ratio was 0.8. Excellent durability on repeated use was observed.[29]

Cobalt coated magnetic iron oxide was dispersed in a glycidyl ether copolymer and applied to a 5 μm polyester film. The film was cured with an electron beam under a nitrogen atmosphere.[30]

Fine iron powder (300 Angstroms) and alumina were mixed in a solution of copolymer of vinyl acetate-chloride in cyclohexane and methyl ethyl ketone. Compositions are suitable for magnetic coatings for disks and tapes.[31]

Magnetic recording media was deposited on a 10 μm polyester film, with a layer thickness of 100 Angstroms of cobalt-nickel alloy vaporized at 5×10^{-5} torr with an electric beam.[32]

Magnetic powder dispersed in silicone oil.[33]

Compressed powdered magnetic cores involving powdered metallics (as from rapid solidification processes) have been described. The amorphous alloy, Fe (72%), Co (8%), Si (5%), and B (15%), is blended with 1% to 2% by weight of an epoxy and 0.5% to 2% of molybdenum disulfide. Particle size averaged 0.7 μm. Parts were molded under heat and pressure.[34]

Plastic magnets were prepared from rare earth magnetizable powders of cobalt-samarium. A composite was prepared of polymer/metals from 61% of powdered poly (tetramethylene terephthalate) and 39% of cobalt-samarium. The units were magnetized 6 seconds at 30 amperes direct current. Maximum energy produced was 5.4 gauss oersteds $\times 10^6$.[35] In another example of a polymer-rare earth metal powder molded into permanent magnets, an epoxy binder was used. The metallic powder which occupied 78% of the volume, was ground earlier to a 20 μm size. Parts were cured 2 hours at 150°C.[36]

Glass fiber reinforced magnetic plastic sheets were used in the manufacture of servomotors and actuators. Magnetizable metal particles consisted of $SrOFe_{12}O_{18}$-silver treated particles. In a diallyl phthalate polymer system, a 2 mm magnetic sheet was extruded and laminated on both sides with a thin nonwoven glass fabric (20 g per m²) for added strength. Tensile strength was 36 kg/mm² and flux density 700 to 750 gausses.[37]

Improved magnetic properties are realized from electrodeposited iron powders, which were protected from oxidation by a toluene solution of an epoxy oligomer *covering* the electroplating bath. Electrodeposition took place from an aqueous $FeCl_2$ electrolyte at 1.8 ka/m² at 318°K.[38] This reference suggests the use of protective layers in electroplating baths, which in some instances could be nonreactive petroleum base derivatives.

MAGNET WIRE ENAMELS

Magnet wire is an electrical metal conductor (usually copper) that is employed to produce or maintain a useful magnetic field. Enameled magnet wire is coated with a smooth, comparatively thin film which is intended to electrically insulate adjacent turns from each other in a winding coil. Magnet wires are used principally in the windings of electrical apparatus such as motors, transformers, solenoids, and control devices. In Chapter 6 a cross section of a motor field coil was shown as a large scale coil in which supplementary polymer insulation was impregnated into the structure to displace any voids.

In the applications to micro- and nano-electronic devices, magnet wires are replaced by layers of insulation and conductive channels associated with semiconductors and printed circuits.

The typical enamel (polymer) coatings are produced by successive applications of enamel, which is baked, until the desired thickness is obtained. The thinner coatings appear on the smallest sizes. The magnet wires are classified according to their thermal endurance. The polymer enamels for commercial classes are listed in Table 8-5 (Refer to NEMA Magnet Wire Specs-MW-1000-1981.)

Almost all magnet wire is insulated soft drawn electrolyte copper wire. The increase in wire diameter from a single coat of enamel varies from about 1 mil (25 μm) for a 2 mm diameter wire to about 5 mils (125 μm) for a wire diameter of 5 mm.

Nylon coated polyethylene is widely used as an insulating coating for twisted wire pairs because of its low cost. It melts at relatively low levels of thermal radiation. Research has been carried out on new materials including radiation crosslinked ethylene-tetrafluoroethylene, Teflon (PTFE) and Kapton (polypyromellitimide) on glass tape, polyalkenes (PK), and polyvinylidene fluoride (PVF$_2$). Most organics charred at levels between 2.1 and 2.9 MJ/m^2 (50 to 70 cal/cm^2), and color coding of the wire insulation underwent change.[39] Jacket thicknesses were principally 0.2 to 0.4 mm.

In developing an improved telephone drop wire for servicing homes from outside transmission lines, a procedure was established for improving the adhesion of polyvinyl chloride polymer (PVC), to the metal core functioning as the electrical conductor. In lieu of extruding a plasticized PVC sheath over the conductor, the following procedure was used. A metal conductor consisting of a steel core covered by a copper sheath was heated to between 330° to 360°C. Discrete particles of unplasticized polyvinyl chloride were deposited on it. On the immediate degradation of the PVC, a discontinuous, very tightly adherent

Table 8-5 Magnet Wire Insulation

	TEMPERATURE RATING (°C)
Class A: Acrylic copolymers, polyamides (Nylon 6-6), polyurethanes	105
Class B: Epoxy resins and glass fibers with heat resistant varnishes	130
Class F: Polyesters	155
Class H: Polyimides and Teflon	180

coating was formed. A plasticized PVC resin coating was extruded over the metal conductors and the tightly bonded discrete adhesion sites. In service, this polymer/metal composite provided an acceptable insulated wire for telephone drop lines.[40]

REFERENCES

1. *Time Magazine*, p. 59, (August 22, 1988).
2. "Magnetic Materials," *McGraw Hill Encyclopedia of Science and Technology*, Vol. 10, (1987).
3. Chin, G.Y. *Science*, 208:888, (May 23, 1980).
4. Kumar, K., *J1 of Applied Physics 63 (6):* R-13, (March 15, 1988).
5. LNP Corp., *Molded Magnetics*, Plastics Design Forum, (Jan./Feb. 1981).
6. Yamashita, S., et al., *32d SAMPE Conference*, 32:80, (April 1987).
7. Delmonte, J., *Metal Filled Plastics*, p. 157, New York: Reinhold Publishing Co., (1961).
8. Robinson, A., *Science*, 233:922, (March 2, 1944) and *Applied Physics Letters*, 46 (8), (April 15, 1984).
9. Bradshaw, R., to IBM Corp., U.S. 4,525,424, (June 25, 1985).
10. Krenceski, M., *Proc. of Polymer Materials*, A.C.S.-57 (463), Fall 1987.
11. Coltman, J., "The Transformer" 258(1) *Scientific American*, (Jan. 1988).
12. Belden, R., *Proceedings of International Conference on Rapidly Solidified Materials*, p. 379, A.S.M. and A.I.M.E. (Feb. 1985).
13. *Proceedings of 31st Annual Conference on Magnetism* Edited by N. Koon, et al., *Journal of Applied Physics*, 61 (8), Part II, (1987).
14. U.S. Patent 874,908, Claims 193-199, (Dec. 24, 1907).
15. White, R., *Science* 229:11 (July 5, 1985).
16. Sony Corp., U. S. 4,210,946, (July 1, 1980).
17. Arnoldussen, T. and Rossi, E., (I.B.M.), *Ann. Rev. Matl. Sci.,* 15:379, (1985).
18. Burguette, M. and Foss, G., (Minnesota Mining & Mfg.), U.S. 4,526,833, (July 2, 1985).
19. Hall, M., *New Scientist*, p. 41, (Feb. 5, 1987).
20. Galli, E., *Plastics Compounding*, p. 27, (Sept./Oct. 1987).
21. Kerekes, Z., *Handbook of Fillers*, ch. 12, p. 205, H. Katz and J. Milewski, New York: Van Nostrand Reinhold Co., (1978).
22. Metzer, N., *Mosaic* 17 (3):29, (Fall 1986).
23. Geiss, E., *Science* 208:938, (May 23, 1980).
24. Wittman, M., *Plastics Engineering*, 43:37, (Dec. 1987).
25. C.A. 101-112348e, Teijin Ltd., J.P. 59,112,018, (June 28, 1984).
26. C.A. 101-115420H, Toyo Kogyo Ltd., J.P. 59,83,702, (May 15, 1984).
27. C.A. 104-131636a, Konishi Roko Photo, J.P. 60,179,927, (Sept. 13, 1985).
28. C.A. 105-154845c, TDK Corp., U.S. 4,600,650, (July 15, 1986).
29. C.A. 105-154830u, Canon K.K., J.P. 61-76,573, (April 19, 1986).
30. C.A. 105-154831v, Nippon Zenn Co., J.P. 61-89,207, (May 7, 1986).
31. C.A. 100-184640w, Fuji Photo, J.P. 58,200,423, (Nov. 22, 1983).
32. C.A. 100-160721x, Matsushita Electric, J.P. 58,189,832, (Nov. 5, 1983).
33. C.A. 106-112474w, Toray Silicon Co., J.P. 61,225,806, (Oct. 6, 1986).
34. C.A. 106-112482x, Fuji Electric Co., J.P. 61,251,108, (Nov. 8, 1986).
35. C.A. 106-157630k, (May 18, 1987), Mitsubishi Rayon, J.P. 61,250,052, (Nov. 7, 1986).

36. C.A. 106(18) 139543E, Seiko-Epson Co., J.P. 61,272,914, (Dec. 3, 1986).
37. C.A. 106(18) 139539H, Matsushita Electric Co., J.P. 61,263,741, (Nov. 21, 1986).
38. C.A. 100-143009d, Poroshk Metall (Kiev), Vol. 2, p. 6, (1984).
39. Schallhorn, B., et al., *International SAMPE Proc.*, 32:1,232, (April 1987).
40. Lerarsky, K., et al., AT&T, U.S. 4,610,909, (Sept. 9, 1986).
41. Ryan, T., (Imp. Chem. Ind.), U.S. 4,708,976, (Nov. 24, 1987).
42. Bowmer, T., et al., *A.C.S. Polymer Preprints*, Los Angeles, 29 (2):258, (Sept. 1988).
43. Nakamae, K., et al., *A.C.S. Polymer Preprints*, Los Angeles, 29 (2):268, (Sept. 1988).
44. Hu, Y. and Talke, F., *A.C.S. Polymer Preprints*, Los Angeles, 29 (2):277, (Sept. 1988).
45. Jones, R.S. and Kuder, J., "Organic Material in Optical Storage," *A.C.S. Polymer Preprints*, Los Angeles, 29 (2):195–223. (Sept. 1988).

9

MICRO AND NANO ELECTRONIC APPLICATIONS

The examination of the use of metal/polymer composites in the electronic industry requires a re-examination of the expressions used in earlier chapters. The composites to be reviewed will be characterized by the contiguous association of two or more diverse materials which leads to a synergistic or superior performance. In large dimension composites such as oriented fiber reinforced polymers which are used in advanced structures, it is not difficult to visualize the roles played by the high strength fibers in the relatively low strength matrix body. The electrical or electronic devices associated with semiconductors, photovoltaic devices, electrical energy distribution, the utilization of magnetic particles on disks and tapes, and the small energy storage capabilities of batteries and capacitors, may entail the use of thin foils, adhesives, bonding elements, and insulative surfaces that are in the micrometer and submicrometer range. The fact that these devices can not always be visually identified as composites does not exclude them from the sophisticated arena of advanced composites. In fact, as molecular dimensions are approached, the rheology of material combinations and associative charge transfer phenomena do not necessarily follow the patterns of the macroscale processes.

In this chapter, the electrical and electronic devices to be considered will be examined for the contributions of the metal/polymer composites. These examinations require research at the molecular and atomic levels. The subjects are introduced to identify the major roles of materials (polymers, metals, and ceramics) in the expanding fields of composites. Remarkable discoveries at the submicroscopic level are contributing to an age of composites. Among the subjects to be covered in this chapter are the following, each of which could be the subject of several volumes:

1. Semiconductor materials, microchips, and thin films,
2. Integrated circuits, and micrographic reproductions,
3. Printed circuit boards (PCBs) and their quality control,
4. Packaging and electrical interconnections of microcomponents,
5. Storage batteries, capacitors, and resistors,
6. Photovoltaic devices, and solar cells.

Details on electroconductive materials, including molding compounds, and coating materials have been discussed in earlier chapters. The attenuation and transmission characteristics of thin gold and copper films, pertinent to semiconductor devices, are discussed in the chapter on coatings. Similarly, attention has been given to ion-implantation techniques which are spawning a new generation of composites for semiconductive devices.

Metal films are applied in micrometer dimensions or less by vacuum deposition, cathode sputtering, plasma arc techniques, ion-implantation, chemical vapor deposition, and electroplating methods. The ion-implantation techniques make available new procedures for directing element distribution into specific areas. This is a basic requirement of many semiconductor devices.

SEMICONDUCTORS

The subject of semiconductors appeared in Chapter 4, where materials with a broad spectrum of electrical conductivities were discussed. Metal filled polymers and electroactive polymers doped with charge-transfer salts were reviewed. In addition to relative comparisons of electrical conductivity (and volumetric resistivity), the temperature dependencies of metals were compared with those of semiconductors. The roles of semiconductors in electronic devices underscore the proliferation of electronic composites of metals, ceramics, and polymers. Visual identifications are no longer possible, but the evidence of millions of electronic devices speak eloquently of their success. Figure 9-1 shows a bipolar semiconductor device from IBM, depicting the functional distributions of metallic semiconductors and the complex interactions which occur in this component.

Because the numbers of experimental semiconductors are growing, some limitations to this discussion are necessary. These limitations are arbitrarily established by calling attention to the principal semiconductors used in electronics as seen in the property comparisons of Table 9-1, silicon (Si), gallium arsenide (GaAs), germanium (Ge), and silicon dioxide, the last, a frequent insulative component, are included. The suitability of a material for semiconductor applications is usually associated with the energy gap between the lowest conduction band and the highest valence band. Greater electronic mobility occurs when lower gap energy must be overcome. Elemental semiconductors include silicon, germanium, selenium, tellurium, and boron. A large number of semiconducting compounds such as copper oxide, zinc sulfide, zinc selenide, indium antimonide, cadmium sulfide, gallium arsenide, lead selenide, lead sulfide, and mercury cadmium telluride, etc., are known and have been used in commercial products. The properties of semiconductors are very sensitive to the presence of impurities, crystal imperfections, and external influences such as temperature, pressure, and the frequency of applied fields.

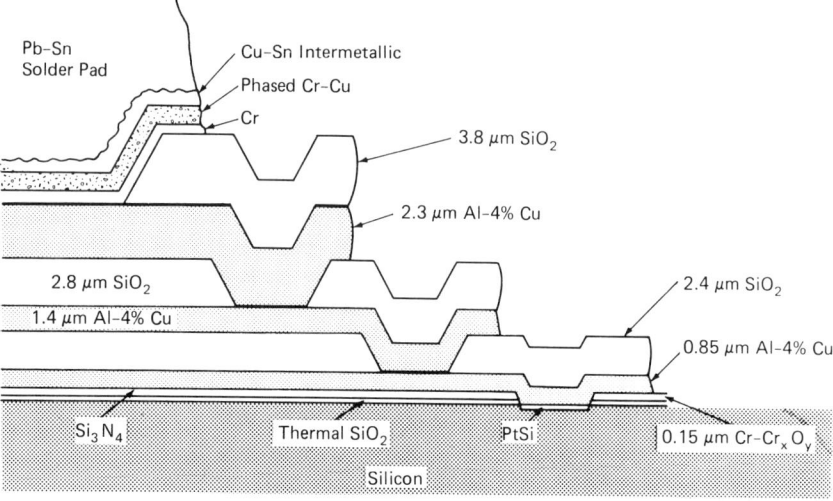

Figure 9-1. Bipolar semiconductor by IBM showing the microscopic composites which are present. Courtesy of IBM Corporation.

Table 9-1 Comparison of Semiconductor Materials[2, 6]

	GERMANIUM	SILICON	GALLIUM ARSENIDE	SILICON DIOXIDE[*]
Atomic weight	72.6	28.09	144.63	60.08
Density-g/cc	5.32	2.33	5.65	2.27
Energy gap[#] (Electron volts)	.80	1.12	1.43	8.0
Dielectric constant	16.0	11.7	10.9	3.75
Linear coefficient of thermal expansion per °C × 10⁻⁶	5.8	2.5	5.9	0.5

[*]Silicon Dioxide substrate appears frequently with semiconductors.
[#]Separation between energy of lowest conduction band and highest valence band is called Energy Band Gap.

The introduction of impurities, called dopants, is accomplished by adding the desired quantity of the impurity to the melt from which the crystal is grown. Ion implantation, whereby highly accelerated ions are impinged on the surface is another technique for the introduction of impurities. The solid may require high temperature annealing to diffuse the implanted ions to the desired locations in the crystal lattice.[1] This may be accomplished by a short burst of laser heating. There are two classes of semiconductors prepared from silicon and germanium:

1. n-type. Excess negatively charged carriers. Typical dopants are arsenic, phosphorus, and antimony.
2. p-type. Excess of positively charged carriers. Typical dopant is boron.

Metals that are semiconductors (group IV) are elements that have four electrons in their outer ring, such as germanium and silicon; or compounds of metals from groups III and V of the atomic table, such as gallium arsenide and gallium arsenide phosphide, the more common compound semiconductors.

If a trivalent impurity (dopant) is introduced into silicon or germanium, holes are created, and the material is said to be p-type. A pentavalent element creates free electrons and the material is classed as an n-type. The quantity of dopant added to create the desired conductivity is much less than 0.1%.[2] The atoms of the crystal lattice are in close proximity, and the electron orbitals of different atoms tend to establish well separated energy levels.

Whether a given sample of a semiconductor is n- or p-type, can be determined by the Hall Effect. If an electric current flows through a sample of semiconductor material and a magnetic field is applied in a direction perpendicular to the current, the charge carriers are crowded to one side of the sample. The transverse electric field is known as the Hall Effect.

The manufacture and preparation of microchips for electronic components account for a six billion dollar industry. Japanese semiconductor companies, represented by NEC of Japan, Toshiba, and Hitachi attained almost 50% of the world chip market in 1987 (*Los Angeles Times,* January 5, 1988). The top five companies in the United States are Motorola, Texas Instruments, Intel, National Semiconductor, and Advanced Micro-Devices (representing 39% of the world market). The diverse roles of microchips in the electronic industry are primarily fulfilled as ceramic-metal composites. The sophisticated and complex composites that are formed are exemplified by the sketch of Figure 9-1. Insulative areas are served in many instances by silicon dioxide and high temperature resistant polymers.

Silicon has been the most widely used material in the semiconductor industry.[8] Gallium arsenide (GaAs) is also suitable for making integrated circuits, transistors, and a host of other microelectronic devices, although it is more expensive. It is estimated that silicon semiconductors account for up to 90% of the semiconductive devices. Both silicon and gallium arsenide are photosensitive and convert incoming light into electrical energy. Gallium arsenide is much more resistant to powerful radiation than is silicon. This is a major reason why it is used in satellite receivers, transmitters, and solar panels. While the limit for frequency is about 4 GHz for silicon amplifier circuits, gallium arsenide amplifiers are good to at least 30 GHz. Gallium arsenide has a much higher electron mobility than silicon. The major technical drawback to GaAs semiconductors is its high sensitivity to traces of element impurities. Costs rise significantly as more stringent purification measures are adopted for GaAs.[3]

The current research on thin films and metal interfaces reflects a timely interest in basic science interactions. Good adhesion of materials to one another is very important under thermal cycling. Because the detailed structures of interfaces and thin films can be manipulated with greater control than the structure of bulk solids, attractive techniques have become available with which to study thin films and interfaces. Thin films form systems whose interior is influenced by the proximity of its interfaces. Current analytical techniques used to study interfaces are listed in Reference 4. The heat of bonding radicals to the substrates is also important for gaining knowledge of interactions. Some results, using the differential scanning calorimeter, appear in Chapter 3. Data on the mobility of deposited thin metallic layers in small microscopic crystallites has been published.[5]

Silicon. Silicon is the most abundant electropositive element in the Earth's crust. For electronic or semiconductor grades a purified silicon is essential. Typically this is obtained by hydrogen reduction and thermal decomposition of silicon containing gases such as silicon tetrachloride. It crystallizes in the diamond lattice, has a specific gravity of 2.42 at 20°C, melts at 1420°C, and boils at 3280°C. Semiconductivity in pure silicon is greatly increased by the addition of minute amounts of impurities. Silicon oxidizes rapidly at room temperature to form a protective layer of silicon dioxide (SiO_2) about 1 nm thick. The oxide layer is amorphous up to 1200°C, above that temperature it is crystalline. Silicon semiconductors are generally protected with an oxidized layer formed at 1100° to 1300°C. Because the indexes of expansion of the silicon dioxide and silicon are similar, high temperature processing can be accomplished without warping.

The early vacuum tubes were superceded in 1947 when Bell Laboratories demonstrated a solid state device capable of the amplification of an electric current. This discovery started the microelectronics revolution and the first device was called a transistor. Texas Instruments Company introduced the first silicon transistor in 1954 and silicon/silicon dioxide systems opened the era of semiconductor device miniaturization.[2] The basic processing of silicon wafers is schematically shown in Figure 9-2. The use of photoresists is illustrated to show pattern development.

Single function discrete devices made of semiconductor materials and integrated circuits containing two or more components on a single piece of semi-

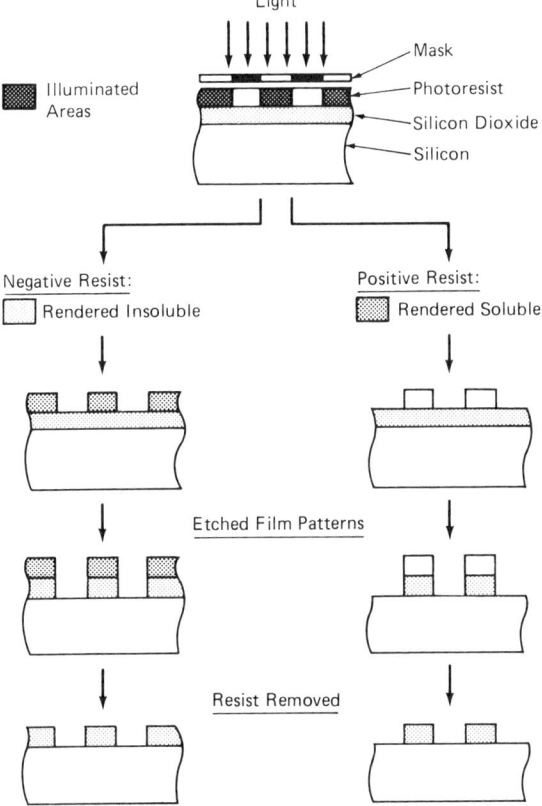

Figure 9-2. Key steps in the fabrication of silicon integrated circuits using photoresists.

conducting material are still being produced today, 25 years after they were first introduced.

The art of growing silicon crystals has evolved into a sophisticated system in which polysilicon and dopant are melted in a graphite crucible by radio frequency heating. The melting temperature is about 1425°C and a seed crystal of the desired lattice structure is inserted into the molten mass (Fig. 9-3). As the seed crystal is slowly withdrawn, a small amount of liquid silicon rises with the seed because of surface tension. When the liquid cools it replicates the crystalline structure. This is known as the Czochralski System (CZ) and is used for the majority of silicon crystals grown for semiconductor use. Crystals up to 4 feet long and 6 inches in diameter are produced, and thin silicon wafers 20 mil (500 μm) thick are sliced from the rod. Quality control procedures are

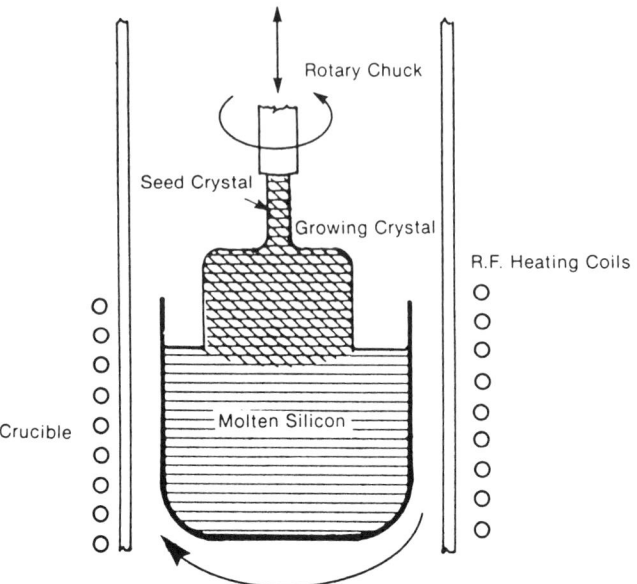

Figure 9-3. Czochralski crystal growing system. Reproduced with the permission of Semiconductor Services, San Jose, CA, and Peter van Zant.[2]

carefully practiced to eliminate defective wafers. Semiconductor device preparations start with these wafers. The wafer surface is polished smooth, preferably so that height of its surface variations fall within 10 Angstrom units.

When being processed into an integrated circuit, the wafer is heated in the presence of oxygen to form a thin layer of silicon dioxide. In subsequent photoresist and microlithography processes, the multiple transistors are reproduced on individual chips.

Table 9-2 summarizes the trend towards miniaturization now taking place with up to 100,000 transistors on a semiconducting silicon chip. Up to 16

Table 9-2 Miniaturization of Electronic Devices[9]

DEVICE INTEGRATION	NUMBER OF CIRCUITS PER CHIP	SMALLEST FEATURE SIZE (μM)
Small-scale integration (SSI)	up to 100	10.0
Large-scale integration (LSI)	10^3 to 10^4	3.0
Very large-scale integration (VLSI)	more than 10^4	less than 1.0

million bits of information are capable of being stored on these chips. (A bit, or binary digit, is the fundamental unit of computer information corresponding to a choice between yes or no.).[9] Earlier semiconductor circuits used 700 to 900 chips per silicon wafer, while the recent increased density of circuits requires larger chips and 200 to 300 chips per silicon wafer have proved to be feasible. A fine diamond surfaced cutting tool separates the chips on the silicon wafer.

Since the advent of the integrated circuit in 1959, the number of transistors that have been placed on a silicon chip has reached one million. Plans for four million and sixteen million chips are underway. Meindl points out that random statistical fluctuations in the energy levels of atoms and electrons in the crystalline structure must not exceed the signals being generated in the device.[43] The properties of silicon establish its unique position in integrated circuits. Silicon's low energy band gap (Table 9–1) maintains excellent semiconductor properties over a temperature range of about 27°C. Meindl shows a silicon wafer containing 470 computer chips. The chips are cut apart with a diamond-tipped saw. Each chip, approximately 6 × 4 mm, contains the central processing unit of a minicomputer.

For extremely small devices, the advantages of very small silicon chips over gallium arsenide, lie in the better thermal conductivity of silicon (approximately a 3-fold advantage). This factor will be advantageous in the higher power concentrations that arise.[43]

Gallium Arsenide. Gallium arsenide semiconductor devices have been the subject of numerous patents by the Toshiba Company. Among the features mentioned in the patents were Schottky barrier gate electrodes on GaAs substrates; the implanting of impurity ions into substrates using gate electrodes as a mask; and the forming of source and drain regions by heat activation.[7]

Gallium arsenide single crystals may be produced by the Czochralski method of using seed crystals drawing cylinders from the GaAs melt (at 1238°C). The arsenic vapor pressure over the melt must be closely maintained during solidification. Because of the presence of two elements, GaAs crystal growth presents more complications than does traditional silicon growth. The wafers that are used are quite fragile and subject to fabrication damage.[11] Gallium arsenide is suited for making lasers for improved optical fiber and satellite communications according to Vander Veen. This semiconductor is resistant to solar and cosmic radiation and is more satisfactory than silicon.

The lasers have been fabricated about a sandwich structure of GaAs and GaAlAs. The GaAs is positioned between thin GaAlAs films creating an efficient laser that emits light at wavelengths of 0.75 to 0.88 μm. Thin layers are grown epitaxially to produce the sandwich layers essential for lasers. Molecular beam epitaxy is the preferred method as it minimizes damage by operating in an ultra high vacuum with control over contaminant levels. High quality sandwich layers, .05 μm thick are grown on top of patterned substrates.

Ion Implantation. Ion implantation has become of industrial importance for the fabrication of semiconductor integrated circuits. By varying the energy and the doses of ion implantation, specific characteristics may be imparted to the substrate. For silicon, it is advantageous to form an amorphous layer during implantation. This layer may be regrown epitaxially, well below the melting point of the silicon. Most commercial implanters have been limited to terminal voltages under 200 kV. Boron, arsenic and phosphorus ions are most important for p-type terminals. With GaAs, the major ion implantation is used for the production of n-type layers. Following ion implants, samples are annealed rapidly at high temperature to remove radiation damage and promote activation of implanted ions.[12]

Ion implantation can add new dimensions to material studies. A number of chemical elements have been introduced into a target substrate by ion bombardment at concentrations considerably greater than normal solid solubilities. The interaction of accelerated ions produces lattice defects. Experiments have been directed at understanding the mechanisms of formation of buried oxide and silicide layers with high dose ion implantation. The damaged region can be recrystallized by annealing, and with most dopants at low concentrations, satisfactory regrowth of the substrate is obtained. These techniques have also been used to implant yttrium into multilayer films of barium copper to create a buried superconductor layer.[13]

In an examination of the superconducting properties of $Ba_2YCu_3O_7$, the critical superconducting current density, J_c, as a function of the applied magnetic field at 25°C, was determined for different forms of the superconductor. Because of the necessity for operating superconductors in strong applied magnetic fields, the critical current that can be sustained is one of the crucial tests of their utility. In Figure 9-4, the critical current, J_c, is illustrated for thin films versus a bulk polycrystalline form. The superiority of the thin film performance suggest that technology learned from semiconductors would be helpful in attaining high currents in strong magnetic fields utilizing thin film technology.[14]

The advantages of implantation into surfaces of semiconductors are attained without changing the properties of the underlying semiconductors. In a study of substrate doping by vacuum deposition of 3, 4, 9, 10 perylene tetracarboxylic dianhydride (PTCDA), a comparison was made of the influence of this dopant upon different organic thin film substrates. Table 9-3 shows some of the results.[15] The current flow under reverse bias breakdown voltage is the criteron.

This particular technique is described as a nondestructive test recommended

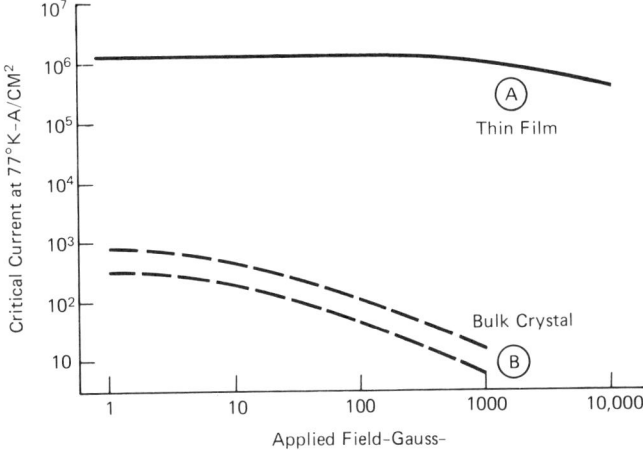

Figure 9-4. The influence of the magnetic field on superconductor critical current. Adapted from Reference 14.

Table 9-3. Influence of (PTCDA) Upon the Surface of Thin Film Semiconductors[15]

SUBSTRATE	DOPING CONCENTRATION PER CC	BREAKDOWN VOLTAGE (V_B)	CURRENT FLOW UNDER ½ REVERSE BREAKDOWN VOLTAGE
pSi	1.5×10^{15}	240	50 microamps/cm^2
nSi	5.5×10^{15}	100	4 milliamps/cm^2
pGe	$5. \times 10^{14}$	170	20 milliamps/cm^2
pGaS	2×10^{16}	10	4 microamps/cm^2
nGaS	3×10^{15}	30	40 microamps/cm^2

for the study of semiconductor wafer surfaces and for assessing the properties of epitaxial films deposited on surfaces. A very thin layer of PTCDA is sublimed on the surface at 500°C. The analysis determines abrupt changes in free carrier concentration at various depths in the epitaxial film (typically 2 to 3 μm) by an organic/inorganic capacitance voltage technique.

Photoresists. Photoresists are important in the lithographic processes developed for the electronic industries. They allow the processes to be accomodated to the decreasing sizes of features to be doped, etched, or plated onto integrated circuit chips. The photoresists are generally polymer films which are applied to semiconductor substrates. They are activated after they are applied as films and have been patterned. The activation exposes areas of the silicon wafers for successive etching, doping, or plating. Exposure on a photoresist surface results in tapering off at edges and less than a sharp slope down to the substrate. Two layer systems now in development involve phenolic resins and polystyrene doped with nitrone and new organometallic resists.[16] Several of the new developments utilize the krypton fluoride excimer laser (248 nm) for lithography.

Polymeric materials for microlithography are able to print dimensions of 0.6 to 0.8 μm. This is accomplished by several step and repeat reductions (5- to 10-fold). For sharpest resolution, excimer laser sources have been used.[18] The basic positive photoresist used in this report was diazonaptho-quinone-novolac. Polymer materials (usually applied by spin coating) which became less soluble after radiation exposure are the negative resists. The positive resists are polymer films that exhibit enhanced solubility after exposure to radiation. Copolymers of glycidyl-methacrylate and ethyl acrylate have good lithographic properties, though polymethyl methacrylate is the most popular for positive resists.

The spin coating technology is influenced by the viscosity of the polymer photoresist and the final spinning speed. Film thicknesses range from 0.5 μm to several micrometers in thickness. To ensure complete and uniform coating an excess of 2 to 5 cm^3 of resist is deposited on a 7.5 cm diameter silicon wafer and spin speeds of 3000 to 9000 RPM are used to develop the film thicknesses. A new class of polyimide siloxane has demonstrated improvements over duPont's polyimide (PI-2545), with uniform films over 4000 to 6000 RPM rotational speed of spin coating.[52] These films are etched by KrF excimer laser.

Electron beams and x-ray lithography are capable of defining features smaller than one micrometer. On May 10, 1988, IBM announced the availability of chip features of a half micrometer (Fig. 9–5). The sizes of circuit elements are continuing to decrease as improvements in lithography and etching occur.

Beam controlled processing technologies for microelectronics have become essential in the fabrication of microelectronics. They are essential to the deposition, lithography, packaging, and related processing of materials used in integrated circuits. Fabrication techniques demand an increasing precision as the physical size of semiconductor devices shrink to submicrometer dimensions.[17]

Figure 9–5 depicts the resolution of 100 nanometers attained at IBM in 1988. The fine resolutions permit greater packaging density, and higher output speeds needed for advanced computers. With accelerating potentials of 20 kV to 200 kV and focusing magnetic fields, electron beams may be converged onto spots a few angstroms in diameter. The focused electron beam will permit semiconductor device developments of remarkable productivity.

Lithography can be divided into writing and printing. For complicated pat-

MICRO AND NANO ELECTRONIC APPLICATIONS 221

Figure 9-5. A scanning electron micrograph of a silicon transistor. Features down to 0.1 micrometer are visible. Courtesy of IBM Corporation.

Table 9-4 Lithography Sources[9,10]

SOURCE	RESOLUTION LIMITS
Mercury arc lamp (H-line—.437 μm)	800 nanometers
KrF excimer laser (Wavelength—248 nm)	350 nanometers
Focused electron beam*	100 nanometers

*Experimental

terns, printing is much faster than writing and is preferred for manufacturing. The dominant printing technology in the microelectronics industry uses patterns of light to change the solubility of the polymeric materials which are used as photoresists. The resolution of this process approaches one micrometer, and is limited by the wavelength of light.[10] At shorter wavelengths x-rays can be used for lithography, but for the nanostructures, electron printing is used. The pattern is usually defined, with polymethyl methacrylate photoresist film, on a rigid quartz substrate. Under illumination, electrons are emitted from the active part of the pattern and focused. Good alignment and accuracy have been attained.

INTEGRATED CIRCUITS

Integrated circuits consist of a combination of active electronic devices, such as transistors and diodes, with passive components, such as resistors and capacitors, within and upon a single semiconductor crystal, such as silicon. Electrically active impurities are introduced into well defined regions of the semiconductor. The photoresist polymer films are cured to define areas which are to be treated, or removed as previously described. Microchip processes are designed to fabricate the transistor structure which is the heart of the circuit. Integrated circuits can be classified into two groups on the basis of the type of transistors they employ: bipolar integrated circuits, in which the principal element is the bipolar junction transistor, and the metal oxide semiconductor (MOS) transistor. The MOS devices are increasingly popular and are widely used. MOS devices are used in switched capacitor filters for large scale integration, and bipolar circuits are used where the highest logic speed is needed.[8] Integrated circuits based on GaAs have the advantage of fast switching speed. They are more costly than the silicon devices.

Miniature electronic circuits are usually produced within and upon a single silicon crystal. Integrated circuits vary in complexity from the single logic circuit and amplifier (about 1.3 mm^2) to large scale integrated circuits containing hundreds of thousands of transistors and other components (Table 9–2). These provide computer memory circuits and complex logic subsystems. Large scale integrated circuits are used in pocket calculators and electronic watches. Microcomputers have contributed to the spread of computer technology to instruments used in many fields, business machines, and automobiles. The functioning of these transistors depends upon microlithography which establishes the minute circuit geometries on the surface of the silicon wafers, while ion implantation techniques imbed components within the wafer.

The starting material for a bipolar transistor is a single slice of silicon crystal, which may be up to 150 millimeters in diameter and several millimeters thick. Upon this base many chips are formed simultaneously through proprietary processes and techniques. Typically the material is doped with p-type impurities. A thin (25 μm or less) epitaxial film is formed upon the p-type base from n-type impurities within the silicon gas (SiH_4) which is introduced. The silicon slice is placed in an oxygen atmosphere at 1200°C, thus forming a silicon oxide layer impervious to electrically active impurities. Many additional steps are pursued in the fabrication of electrical devices, such as the opening of passages at selected areas of the protective oxide and the exposing of the wafer slice to appropriate dopants. As the features on the chip become smaller, there is an increasing need for ion implantation. In the process, aluminum, for example, would be deposited in a vacuum (area control by masks), and the excess would be removed by photoengraving, leaving behind aluminum stripes, or connectors, and resistors.[2,8] In the MOS fabrication process, the incoming wafers are processed through oxidation and go directly to masking. MOS, with no need for the isolation structure of bipolar technology, allows a higher component density.

The integrated circuit fabrication process must be guarded at all times from impurities of all description and precise cleanliness is required in the manufacturing area. Gallium arsenide integrated circuits have the advantage of fast switching speed. As with silicon wafers, precise care in handling and cleanliness are essential. The gate of the GaAs field effect transistor controls the path of the carrier flow by means of the potential applied to the gate. The gate is a Schottky barrier composed of metal and gallium oxide. While this overview of integrated circuits is necessarily limited, it does convey the extensive interplay of metals, polymers, and ceramics at the micrometer level and below. These are the nonvisible examples of electronic devices which will open the new era of composites.

PACKAGING FOR ELECTRONICS

Developments in semiconductor based products have been accelerating because of advances in chip technology. These developments have been reflected in the number of devices per chip and the substantial increase of switching speed, as compared with older counterparts. The manner in which semiconductor chips are connected to form a semiconductor-based-electronic-unit is defined as electronic packaging.[19] Hofer discusses electronic packaging at three levels.

Chip Packaging. This refers to the mounting of chips (silicon or GaAs) on a carrier so that it may be mounted on a board or module. A transfer molded epoxy, one on a lead frame is shown in Figure 9-6, is a good example of a molded metal/polymer composite for electronics. Ease of molding, rapid cure, and high heat resistance have made these epoxy chip carriers popular in electronic devices.

Board Packaging. Epoxy printed circuit boards (PCB) contain chemically etched or electroplated conductors on one or both sides of a sheet of reinforced epoxy or polyamide. Recent developments involve multiple layers of metal/polymer composites, with precision located electroplated holes used to maintain connections between layers, and to establish greater chip densities.

Traditional printed circuit boards mechanically support, and electrically in-

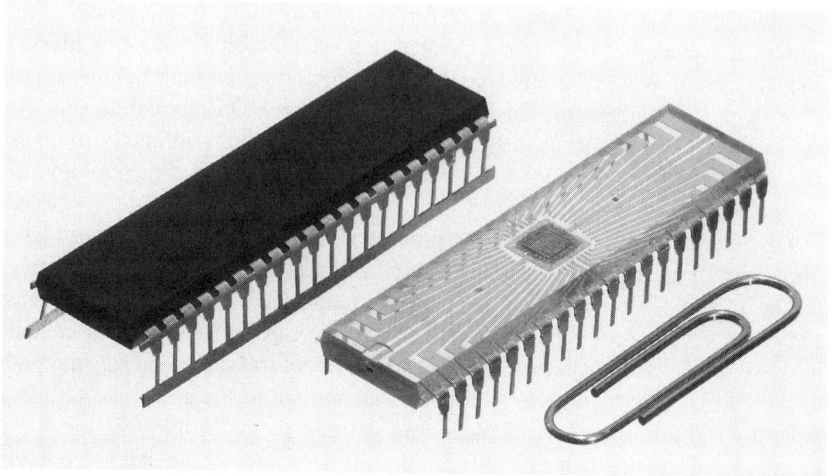

Figure 9-6. Transfer molded epoxy on a chip carrier. Courtesy of Furane Products, Division of Ciba-Geigy.

terconnect components on a two-dimensional surface. Recent developments in moldable plastics have established three-dimensional designs which have expanded the roles of PCBs. A three-dimensional prototype circuit board, such as the one illustrated in Figure 9-7, suggests the possibilities of the multilevel mounting and support of microchips as an alternative procedure for increased device density. The material used was a glass fiber filled polysulfone molding compound.

Features which have been introduced include: three-dimensional additive plating technology; special interconnector molded devices; and through hole plating aspects. Reference 20 describes a two-shot molding procedure wherein a thermoplastic, which is capable of accepting an electroless coating, creates a feature to be metallized. The second injection shot injects a nonplatable plastic amenable to the plating design.

Module Packaging. The module consists of a collection of chips which must function with special consideration for rapid signal mobility and cooling. Electronic devices are being designed with multilevel ceramic/metal construction. Metal pins provide connections to the board. The multilevel designs permit high concentrations of integrated circuits for rapid data processing, and must provide good registration and electrical continuity between several layers and their respective planar surfaces. Both rigid and flexible modules are available. High temperature resistant polyimides are favored among the flexible designs.

MICRO AND NANO ELECTRONIC APPLICATIONS 225

Figure 9-7. A molded circuit board. Courtesy of LNP Corporation.

Molded circuit boards have attracted considerable interest, as they have been directed towards the design objectives of electronic manufacturers. Injection molding of PCBs has been reviewed by LNP Corporation.[21] First introduced in the 1970s with molded thermoplastic polysulfone substrates, they were followed by high performance molded polymers such as polyetherimide, polyarylsulfone, glass filled thermoplastics, and polyether sulfone. With increasing chip density, the molded parts must be able to meet higher operating temperatures and maintain good dimensional stability. Among tests required for rigid printed wiring boards (PWB), including injection molded circuit boards, is a 10 second float in a liquid solder bath at 550°F (275°C) MILP-5505,110D. This temperature exposure will reveal inadequate provisions for differences in thermal expansion coefficients, and poor adhesion of metal/polymer composites (Fig. 9-8).

Microsectioning of printed circuit boards has been recognized as an important quality assurance tool. Current military specifications for PCBs require that test coupons of every multilayer panel be microsectioned in the "as pro-

226 METAL/POLYMER COMPOSITES

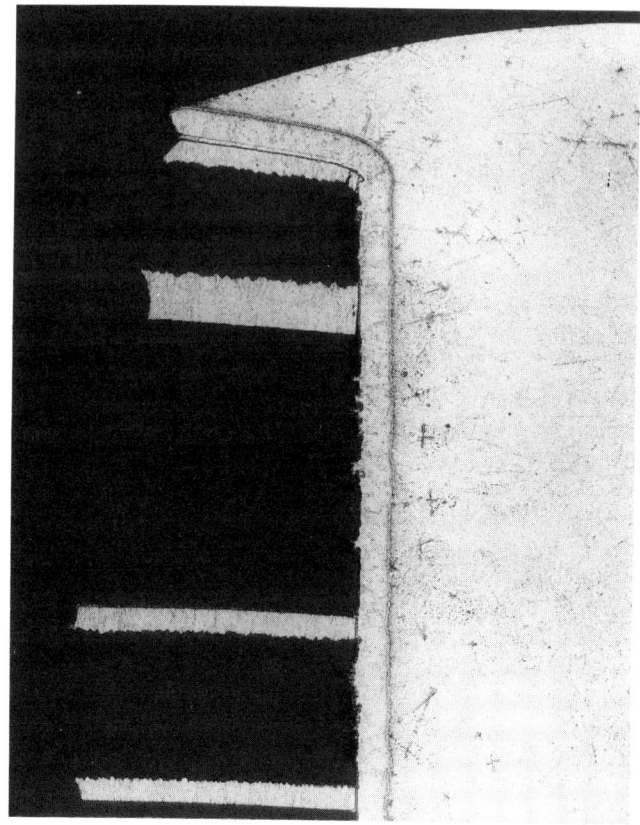

Figure 9-8. A microsection of a multilayer circuit board. Courtesy of Delsen Testing Laboratories, Glendale, CA.

duced'' condition and after thermal stress (solder float) conditioning.[22] This is illustrated in Figure 9-8, which shows a microphotograph of a multilayer board that reveals an internal layer separation, of the 2 mil (50 μm) copper foil, from through-hole plating connection. Defects of through-hole plating are observed, particularly on multilayer boards which may have more than ten layers of copper. Chip devices are mounted on both sides of the board.

There are new developments in composites which will contribute to faster signal speeds on printed circuit boards. For example, unique polyaromatic cyanate esters for PCBs offer low dielectric constants. Signal speed in an insulated conductor is inversely proportional to the square root of the dielectric constant of the insulating material. The thermal properties of Dow's XU-71787 polyaromatic cyanate ester laminates show a higher glass transition temperature, a lower thermal expansion coefficient and a much lower dielectric constant than the epoxy laminates now in use.[53]

The fabrication of multilayer ceramic wiring boards by a new process involving thick films and tape technology permits less costly and faster productions.[54] A glass-ceramic tape layer is transferred to a PCB with an insulating substrate containing a conductive pattern. This replaces multiple printings of thick film dielectric paste. The tape transfer process matches duPont's-4575 thick film process. These materials have expansion coefficients similar to the ceramic chip carriers, and lend themselves to surface mount technology.

Several examples of electronic devices and their materials are listed in the paragraphs below. They are illustrative of printed circuit boards applications and semiconductors.

1. Polyaramid fibers are recommended in paper and fabric form for the reinforcement of flexible or rigid circuit boards. Newly developed aramids from Japan possess fiber diameters of 12 μm, tensile strengths of 310 kg/mm^2, a modulus of elasticity of 7100 kg/mm^2, and a negative thermal coefficient of expansion.[23]

2. Fusible powdered metal pastes with rosin and polyalcohol binders, and solder alloy powders have been developed for microelectronics.[24] Formulations involve modifiers to reduce slumping.

3. For use on liquid crystal displays, transparent electrically conductive films were prepared from 100 μm polyether sulfone film. The film was coated with 5 μm of polyvinyl alcohol and 50 nm of silicon dioxide on which were deposited 25 nm of indium oxide. The units show long service life.[25]

4. Heeger describes metal polymer Schottky barriers on cast films of soluble poly (3 hexylthiophenes). Current voltage rectification ratios are 100 to 1—up to 1000 to 1. Uniform dopant concentration was attained at a depth of 150 nm. Indium was evaporated on the film.[26]

5. Surface-mount chips are growing in popularity to accommodate new chip carrier technologies, and to permit more rapid response by shorter paths between components. Figure 9-9 illustrates a multichip molded carrier. It has been molded with an antistatic compound, a prudent step, used to avoid accumulation of dust particles from the environment. Such particles denigrate performance of the assembly.

6. Substrates for microelectronics have been reviewed by duPont with an emphasis on new materials and processes. Signal propagation velocity for polyimides and silicon were about 0.60 nanoseconds per meter, one-half of their speed in air.[28]

7. A tremendous surge in the filing of patent claims on superconductors has taken place in Japan. Emphasis has been placed on the thin film technique, essential for constructing electronic devices from the new materials. Toshiba

228 METAL/POLYMER COMPOSITES

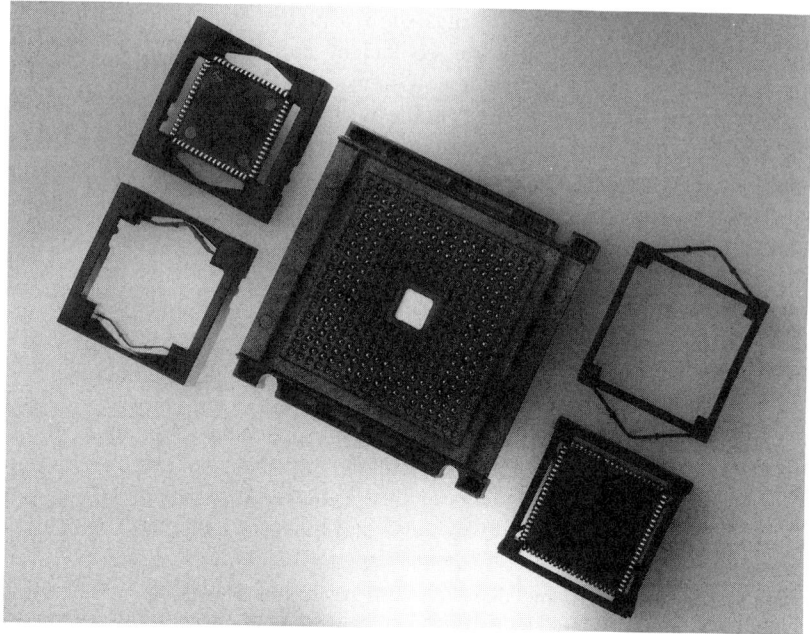

Figure 9-9. A molded chip carrier with statically dissipative nylon. Courtesy of LNP Corporation.

produced superconductors by reactive multitarget sputtering of yttrium, barium, and copper with a mixture of argon and oxygen, forming a thin film 700 nm thick. The surface layer was stabilized with a layer of oxidized silver. Lower processing temperatures of 560°C achieved a smooth surface comparable to today's semiconducting wafers. Paul Chu, an American scientist who pushed superconductivity to the highest temperatures yet achieved, has signed an agreement with duPont for exclusive commercial rights.[38] It is also reported that Japan's industrial giant, Hitachi, has filed more than 600 patents on semiconductors.

STORAGE BATTERIES

Energy storage batteries are prime examples for the application of metal/polymer composites. Introduced nearly a century ago, primary or dry cells have been used to store as much energy as possible into a small container of fixed size. Secondary or rechargeable lead-acid batteries, first described in 1859, are identified with automobiles, providing auxiliary energy for starting ignition, and the lighting of the vehicles. The battery converts energy freed by a chemical reaction directly into electricity.

One of the principal examples of primary batteries is the Leclanche cell, patented over 100 years ago, which is based upon a zinc anode and manganese dioxide. The "dry" cell concept formulated by Carl Gassner of Germany in 1886 greatly increased the popularity of the small battery. The dry cell uses a liquid electrolyte that has been immobilized in a paste or gel. The chemicals, metals, paper, and plastics provide the components for the composite structure. The popular Eveready carbon-zinc dry battery which is widely used in flashlights is illustrated in Figure 9-10. A considerable amount of engineering data has been assembled on these products.[39] An increasingly popular variant, the alkaline cell, which operates at a higher rate, is based upon potassium hydroxide. A core of powdered zinc and alkaline electrolyte serves as the anode mix. The core is separated by a fabric from a cathode mix of manganese dioxide, carbon, and electrolyte and appropriate polymeric insulation.

The alkaline-manganese-dioxide battery has been designed into primary and rechargeable batteries. These products have been available for over 100 years, and one cannot help but speculate on the contributions that can be made from electroactive polymers and new polymers. Metal/polymer composite concepts are adaptable to these areas.

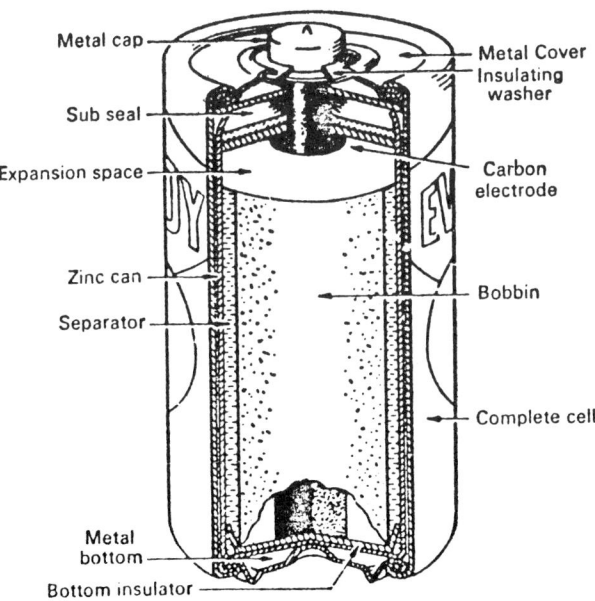

Figure 9-10. A carbon-zinc storage battery. Adapted from Reference 29.

230 METAL/POLYMER COMPOSITES

Mass production of large scale integrated circuits has led to the introduction of electronic watches, calculators and similar appliances, and to the use of primary alkaline cells. These cells possess superior shelf life and performance. Modern batteries such as alkali-zinc-mercuric oxide cells, and zinc-silver oxide cells have good storage characteristics (Fig. 9-11). Lithium batteries also have good storage and discharge characteristics and the capability of operating at low temperatures. Developments in advanced metal/polymer composites will extend the capabilities of primary batteries even further, because of their efficient use of optimum materials.

Rechargeable secondary, lead-acid batteries are undergoing new developments because of manufacturing technology. Cold-worked corrosion resistant wire-reinforced grids are being evaluated by the Navy, and they are expected to double the life of the storage battery.[28] Many thousands of storage battery cases are now being produced from polyolefins as replacement for the semicold molded bitumen pitch compounds. The latter compounds were used extensively in the early part of the twentieth century because of the low cost of bitumen and its good chemical resistance to sulfuric acid. The plastic bases are tougher, lighter in weight, more durable, and are capable of accommodating metal attachments and inserts.

General Motors has focused upon lithium secondary cells for automobiles. With electrodes of metal sulphide and lithium alloys, and electrolytes of LiCl-KCl, 180 Wh/kg have been demonstrated—several times the lead-acid performance.[29] The nickel-cadmium battery is an example of a contemporary rechargeable cell. In addition to regular seals, polyethylene insulators are now

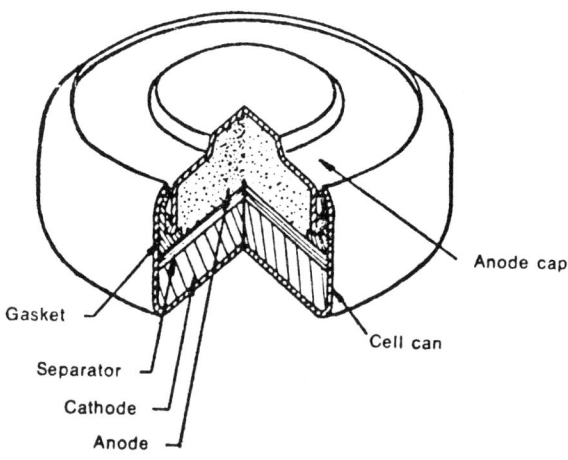

Figure 9-11. A silver-oxide storage battery. Adapted from Reference 29.

used. Coiled electrodes form a cylindrical battery. The positive electrode is nickelic hydroxide (NiOOH), the negative electrode is cadmium, and the electrolyte is an aqueous solution of potassium hydroxide. The Ni-Cd cell has flat voltage discharge characteristics and good storage life.

While domestic energy from electric utilities may be estimated at 5 cents per kilowatt hour (kWh), a D-size Leclanche cell for a handheld flashlight would cost $75 per kWh, while the energy cost for a small zinc silver oxide watch battery is estimated at $10,000 per kWh.[30] Recent developments of note are:

Since the invention of electronically conducting polymers, there have been efforts to develop some of these materials into batteries which would be competitive to lead-acid and Ni-Cd secondary batteries. Low charge density per surface area has handicapped these development efforts.[30]

Separators for electrochemical cells are prepared from microporous films of C_2H_4 and C_2F_4. Lithium was sandwiched between test films in an argon atmosphere sealed in polyethylene bags. Whereas C_2F_4 polymer film blackened in one week at 70°C, copolymers of C_2F_4-C_2H_4 and copolymers of C_2ClF_3 remained clear. Their use in fuel cells may be feasible.[31]

Films of a conductive polymer mix of polypropylene (100 pbw) and carbon black (40 pbw) have been extruded into 100 μm films and roll pressed on a 30 μm foil at 90°C and 80 kg per cm^2 pressure. These composites were used as current collectors for small batteries.[32]

Nickel plated spheres have been developed for alkaline batteries.[33]

Lithium batteries are being used with electroactive polymers in secondary batteries. Thin films of electrolyte are sandwiched between two active electrodes. The thin films must have sufficient ion capacity for the charging and discharging of the electroactive materials. The film must also act as a separator, and in practice this may mean a thickness around 100 μm. Doped polyacetylene and suspensions of polypyrrole have been examined.[34]

A coin rechargeable battery with a polyaniline electrode and a lithium metal electrode went on sale in Japan in 1987. The batteries are said to have a capacity significantly larger than existing rechargeable lithium batteries; the voltage is approximately 9 volts. Prospects are good for long shelf life stability.[42] This battery was developed by Bridgestone Corporation and Seiko Electronics.

West Germany's BASF and Varta Batterie A.G. have developed a 3 volt battery with a polypyrrole film and a lithium electrode. The experimental model is flexible, about 4 mm thick and the size of a post card.[41]

The progress in polymer electrodes offers the prospects for an increasingly visible role for these materials in the manufacture of small batteries for the consumer.

CAPACITORS

Composites of conductors and dielectrics are extensively used in electronic circuits to secure capacitance for the storage of electrical energy in the electric field. The energy stored in a capacitor is:

$$W = \tfrac{1}{2} CV^2 \text{ joules}$$

where C = the capacitance in farads
V is the voltage in volts.

Modern capacitors come in many forms among which are metal foils and dielectrics such as mica, and thin polymer foils such as polystyrene and polyester, the latter have and maintain high insulation resistance. Many electronic devices utilize small scale examples of polymer/metal composites.

Trotter has effectively described the design and manufacture of capacitors, particularly the multilayer ceramic capacitors (MLCCs) used in the design of integrated circuits.[42] Typically they are less than 5.0 mm in size and are often coated with plastic before being soldered to a circuit board. They serve to divert power surges and spurious electrical signals. In the manufacture of the MLCCs, barium titanate (which contributes to high dielectric constants) and other oxide powders are ground to a few micrometers in size and dispersed into a polymer solution and cast into thin films of ceramic tapes. The polymer solvent is evaporated and the electrode patterns printed with a silver-palladium conductive ink. These are formed into stacks of from 30 to 60 sheets, which are diced into thousands of capacitors (Fig. 9–12). Further controlled heating at 1000° to 1400°C burns off the polymer and sinters the ceramic powders into the finished capacitor.

Recent capacitor developments are abstracted below:

1. Al foil (99.5% purity) for capacitors are rolled and annealed at 450° to 600°C, and cooled at a rate of 50°C per minute to under 300°C. The foil is electrolytically etched in HCl or H_3PO_4 to increase its capacity for electrolytic condensers.[35]

2. Polypropylene film with zinc coatings on both sides was developed for rolled capacitor elements. The capacitor was impregnated with epoxidized linseed oil, accompanied by titanate coupling agents.[36] Capacitors were rated at 20 μf.

3. Nonpolar polythiophenylene films are vacuum metallized and used in film capacitors, flexible circuit boards, and magnetic tapes. Heat resistance standards (5 seconds at 250°C) must be met to pass solderability tests. Metal film peel adhesion of this composite withstood 52 g over 5 mm.[37]

4. An anodized aluminum capacitor was manufactured with a TCNQ complex solid electrolyte in combination with N-isoamyl-isoquinolnium and sealed

MICRO AND NANO ELECTRONIC APPLICATIONS   233

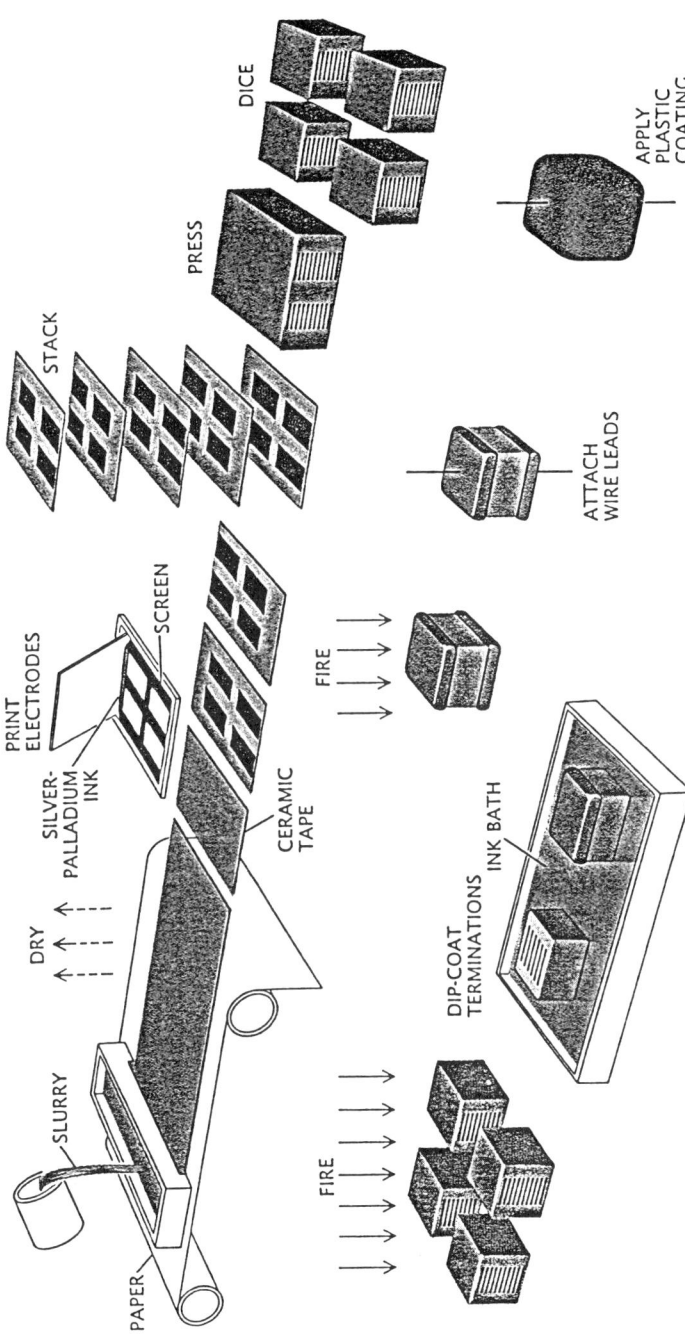

Figure 9-12. The manufacture of multilayer ceramic capacitors. Courtesy of D. Trotter[42] and Scientific American, Inc. Copyright 1988. All rights reserved. Illustration by George Kelvin.

with a heated powdered epoxy resin. The capacitor showed high resistance to reverse currents.[40]

PHOTOVOLTAIC CELLS

The photoelectric current in a photovoltaic cell exposed to the sun is directly proportional to the intensity of the sun's radiation which is strongest about mid-day. Under normal operating conditions the efficiency of conversion of the silicon cell is about 10 to 12%. Measures have been adopted to enhance optical concentration to improve the light efficiency, though higher temperatures must be considered. The spectral response peaks of several photoconducting materials follow:

MATERIAL	PEAK WAVELENGTH (μM)
Cadmium sulfide	0.5
Silicon	0.9
Gallium arsenide	1.3

Varian Associates in Palo Alto, California has developed a single function GaAs solar cell with an efficiency of 28%.[48] These cells are more resistant to sunlight and radiation in space than are silicon cells. The GaAs cell measures 5 × 5 mm, and larger area cells are under development. The generation of electricity to meet peak electrical demands, as in the southwest desert areas of the United States should prove economically popular.

Alternative techniques have used thin film solar cells, which are lower in cost but offer less efficient light conversion, varying from 5 to 9%.[49] Arco Solar, Inc. (California) has developed thin film solar cells with an efficiency of 11.2%. The thin film photovoltaic cell which consists of a film of silicon or other photosensitive film is deposited on a substrate. This contrasts with the single crystal cells described above which are formed of Si and GaS cells. The new thin films formed from copper-indium-diselenide (CIS), appear to be more stable than the thin film silicon cells. However, commercial growth has been slow in early 1989.

Thin films have been deposited by flash evaporation of $CuInSe_2$ powder, by ion-beam sputtering of polycrystalline $CuInSe_2$, and by molecular beam epitaxy.[50] Short circuit currents of 38 ma/cm² were reported with the 12% efficiency. Because the optical absorption coefficient of CIS is very high, the internal photocarrier collection efficiency is near unity, permitting high short circuit currents. The CIS cell requires multiple layers of thin films of several materials to attain optimum efficiency and reliability. Among the composite assemblies were the following assembled on an Al_2O_3 substrate (Boeing Data)[50]:

Components of Boeing Composite Assembly

Conductive Layer of Mo	0.3 μm
Low p-CuInSe$_2$	2.5 μm
High p-CuInSe$_2$	0.8 μm
High p CdS	0.8 μm
Low p CdS	1.7 μm
Radiation-resistant coating	

Polymeric coatings for solar cells must be prepared to provide a high degree of solar radiation protection. There are a few polymer candidates which make effective contributions to metal/ceramic/polymer composites for the solar cell. Polymethyl methacrylate and polyvinyl fluoride films have demonstrated good resistance to discoloration, which, if present, would lessen the efficiency of the cell. Solar protons and electrons, and ultraviolet light contribute to the exposure instability in outer space. An effective polyimide film prepared from 4.4'-hexafluoroisopropylidene bis (phthalic anhydride) dissolved in purified N,N-dimethylacetamide has demonstrated good characteristics in protecting solar cells from radiation. Standard ohm-cm shallow diffused silicon solar cells designated as K6 solar cells were used in the evaluation. The thicknesses of the polyimide layers were in the range of 5 to 12 μm.[47]

Fused quartz covers have been adhesively bonded over silicon cells, however, with silicone adhesives filters were necessary to exclude ultraviolet light which degrades the silicone adhesive. Adhesive bonding of thin ZnO/Ta$_2$O$_5$ coated quartz cover slides over solar cells has provided protection from ultraviolet radiation in other installations.

Silicon solar cell arrays have become the exclusive solar cell power source for satellites. Skylab, launched in 1973, had a 20 kw solar cell array. In the latest development at Sandia National Laboratories (Albuquerque, NM), a composite solar array is prepared with a gallium arsenide cell to respond to blue light, and a silicon bottom cell is prepared to respond to the red end of the spectrum.[48] This solar array has attained improved efficiency. These new composites of metals, ceramics, and polymer coatings are bringing the utilization of solar energy closer to commercial fulfillment. The solar cells developed for space have been too expensive to compete with other methods of electricity generation.[51] Hence research has been directed recently towards thin film techniques. Ultra thin layers of deposited semiconductors, forming single crystal structures, are being routinely fabricated by molecular beam epitaxy.[17]

The solar-power car, the GM Sunraycer, which outperformed competitive vehicles in the 3,000 km race from Darwin to Adelaide, Australia has recently received attention. The race took place during November, 1987. The car's running time was about 45 hours at an average speed of 66.9 km per hour during the 5.25 days involved. The solar cells, the gold plating on the streamlined acrylic canopy, the rechargeable silver zinc cells, and the vehicle components which made use of metal/polymer composites are of special interest, and are discussed below. The lightweight, teardrop streamlined design was produced under the direction of MacCready at AeroVironment (Pasadena, California).[44] Figure 9-13 shows the vehicle. Other solar powered vehicles are illustrated in Reference 45.

Solar Array Fourteen hundred K7 silicon solar cells produced by Hughes Aircraft (of the type used on Hughes-built communication satellites) were produced for the GM Sunraycer. Each cell was 2 by 6 cm, 0.2 mm thick, with a nominal efficiency of 16.5%. In addition, 3,800 gallium arsenide solar cells from Applied Solar Energy Corporation were used. These cells were 2 by 4 cm and 0.3 mm thick, operating with a nominal efficiency of 20%. The solar cells were connected in series and arranged in 20 strings of about 450 cells each. The peak power reached was 1,500 watts at 150 volts within an area of 90 square feet.

Figure 9-13. The GM Sunraycer, powered by 8800 solar cells. Courtesy of G. M. Hughes Electronics.

Batteries. An indirect source of power was the 68 rechargeable silver-zinc cells, each providing 1.5 volts and 25 ampere hours. They weighed a total of 60 pounds, about one-fifth of the weight of a lead acid battery. Battery power was used early and late in the day to supplement the reduced solar power available at those times. Batteries were recharged during the first 2 hours of sunlight.

Motor. Recently developed permanent magnets, made of rare earth-neodymium-boron-iron alloys, were used in the new motor. The motor weighed 17.8 kg and produced 2 horsepower at 4,000 RPM at a high efficiency rate. It was manufactured by Delco-Remy, Division of GM.

Gold-plated Canopy. The gold plating on the transparent acrylic canopy reflected 90% of the visible light and 98% of the infrared radiation. By blocking infrared rays, the canopy helped to keep the driver of the Sunraycer relatively cool.

Body of the Car. The chassis of the GM Sunraycer is a welded aluminum frame weighing 15 pounds, which supports a vehicle weighing 547 pounds. Overall dimensions were 6 m (length), with a width of 2 m, and height of 1 m. The body of the car is a polyaramid honeycomb composite structure which provides good strength and stiffness at a low weight.

References

1. *Van Nostrand's Scientific Encyclopedia,* p. 2,545, New York: Van Nostrand Reinhold Co., (1983).
2. VanZant, P., *Microchips Fabrication,* San Jose, CA: Semiconductor Services, p. 45, (1986).
3. Rogers, D., *ASTM Standardization News,* p. 28, (Oct. 1987).
4. Anon.—*Field of Interfaces and Thin Films* C and E News, 64:22, (Aug. 11, 1986).
5. Bocris, J. et al., p. 330 and Bonzel, H., p. 247, *Structure and Properties of Metal Surfaces,* Tokyo: Maruzen Company, (1973).
6. Liao, S., *Microwave Devices and Circuits,* Englewood Cliffs, NJ: Prentice Hall, (1980).
7. C.A. 106 (20) 167334 to 1 67337, May 18, 1987, Toshiba Corp., J.P. 61,295,666 to 670 (Dec 26, 1986).
8. *McGraw Hill Encyclopedia of Science and Technology,* New York: McGraw Hill, Vol. 16, p. 246, and Vol. 9, p. 243, (1987).
9. Chaudhari, R., *Scientific American,* 255:137, (Oct. 1986).
10. Howard, R.E., et al., *"Nanostructures," Ann. Rev. Matl. Science* 16:441, (1986).
11. Vander Veen, M. R., "GaAs Sandwich Lasers," *Advanced Materials and Processes,* 133:39, (May 1988).
12. Sealy, B. J., *International Materials Review,* 33:38, (1988).
13. White, A. and Short, K., *Science* 241:930, (Aug. 19, 1988).
14. Murphy, D., et al., *Science* 241:922, (Aug. 19, 1988).
15. Forrest, S., et al., "Thin Film Diagnostics," *Ann. Rev. Material Science,* 17:189, (1987).

16. Anon., *Chemical and Engineering News*, p. 27, (Oct., 1985).
17. Kern, D., et al., *Science* 241:936, (Aug. 1988).
18. Reichmannis, E. and Thomas, I., Ann. Rev. Mat. Sci. 6:235, (1976).
19. Hofer, D., *A.C.S. Polymer Reprints* p. 85, (1987).
20. Frisch, D., *Materials Engineering* 105:34, (Aug. 1988).
21. LNP Corp., *Bulletin 400-186*, and Gandre, J., *Printed Circuit Design*, Vol. 3, (June 1986).
22. McCrory, B., "Microsectioning of PCB's," *Electronic Manufacturing*, 34:14, (July 1988).
23. Nishimura, K. and Hirakawa, T., *International SAMPE Conference*, (Anaheim, California), 32:1,200, (April 1987).
24. Hwang, J., (SCM Corp.) U.S. 4,619,715, (Oct. 28, 1986).
25. C.A. 105-135123z, Sumitomo Bakelite, J.P. 61, 7964 and 5, (Apr. 23, 1986).
26. Heeger, A., et al., *Synthetic Metals* 22 (1):63, (1987).
27. Science Applicators Inc. *Tech Report*, La Jolla, CA, (May 1986).
28. "Metal Matrix Composites," *Current Highlights*, (Dec. 1985).
29. *Van Nostrand's Scientific Encyclopedia* p. 316, New York: Van Nostrand Reinhold Co., (1983).
30. Vincent, C., *New Scientist* 101:34, (Mar. 29, 1984). Passiniemi, P. and Osterholm, J. *Synthetic Metals* 18:637, (1987).
31. C.A. 106-105504c, Duracel Inc., U.S. 4,629,666, (Dec. 14, 1986).
32. C.A. 106 (20) 159631k, May 19, 1987, Sumitomo Bakelite, J.P. 62,15,762, (Jan. 24, 1987).
33. Abbaschian, G.A., U.S. 4,565,571, (Jan. 21, 1986).
34. Inganas, O. and Lundstrom, I., *Synthetic Metals*, 21:13, (Aug. 1987).
35. C.A. 108(6) 47742g, Feb 8, 1988, Sumitomo Metals, J.P. 62,193,238, (Aug. 26, 1987).
36. C.A. 106(18)147986s, Matsushita Electric Co., J.P. 61,171,005, (Aug. 1, 1987).
37. C.A. 106-103462p, Matsushita Electric Co., J.P. 61,230,932, (Oct. 15, 1986).
38. Cross, M., *New Scientist*, p. 49, (June 30, 1988) and p. 30, (Sept. 1, 1988).
39. *Eveready Battery Engineering Data*, New York: Union Carbide Battery Products Division, (1976).
40. C.A. 108(6) 47721 to 47724, Matsushita Elec. Co., (Feb. 8, 1988).
41. Kaner, R. and MacDiarmid, A., *Scientific American*, Vol. 258, p. 109, (Feb. 1988).
42. Trotter, D. M., *Scientific American*, Vol. 259, p. 90, (July 1988).
43. Meindl, J., *Scientific American*, Vol. 257, p. 78, (Oct. 1987).
44. MacCready, P., "Sunraycer," *Engineering and Science*, 51:2 Pasadena, CA: California Institute of Technology, (Winter 1988).
45. *Smithsonian Journal*, 18:48, (Feb. 1988).
46. Wigotsky, V., *Plastics Engineering*, p. 21, (Nov. 1987).
47. DuPont, P. and Bilow, N., (Hughes Aircraft), U.S. 4,592,925, (June 3, 1986).
48. Anon., *Chem. Eng. News*, p. 30, (June 6, 1988).
49. Pool, R., *Science* 241:900, (Aug. 19, 1988).
50. Mitchell, K., *Ann. Rev. Matls. Sci.* 12:401, (1982) and 18:4,365, (1977).
51. Perez-Albuerne, E. and Tyan, Y., "Photovoltaic Materials" *Science*, 208, p. 902 (May 23, 1980).
52. Lee, C. et al., "Microelectronic Packaging," *SAMPE Jl.* 24 (March/April 1988).
53. Bogan, G. et al., *SAMPE Jl.* 24:19 (Nov./Dec. 1988).
54. Vitriol, W. and Brown, R. (Hughes Aircraft) U.S. 4,645,552 (Feb. 24, 1987).

GLOSSARY

TERMS USED IN METAL POWDER AND METAL PROCESSING

Definitions of terms frequently used in metal powder terminology, and which are pertinent to the material covered in this book are listed below (adapted in part from ASTM-B-243-76).

Amorphous state. Noncrystalline form of a metal obtained during rapid cooling from the molten state.

Atomized metal powder. Usually produced by the dispersion of a molten metal by a rapidly moving gas or liquid stream.

Carbonyl powder. A metal powder prepared by the thermal decomposition of metal carbonyl.

Cathode sputtering. The high voltage discharge between electrodes propelling metal towards the object to be plated, which functions as an electrode.

Chemical precipitated metal powder. Produced by the reduction of a metal from a solution of its salts either by the addition of another metal higher in the electromotive series or by other reducing agents.

Compact. An object produced by the compression of a metal powder, generally while confined in a die, with or without nonmetallic additives.

Curie temperature. A temperature above which the material's ferromagnetism disappears in a material—usually lower than melting point.

Dendritic powder particles. Usually of electrolytic origin, having a typical pine tree structure.

Densification. A term associated with compaction processes to minimize voids.

Electroless copper or nickel coatings. Proprietary compounds for depositing thin films of metal to establish electroconductive surfaces. Further build up may take place by electroplating under a direct current voltage.

Electrolytic powder. Produced by electrolytic deposition of metallic ions or by pulverizing an electrodeposited metal.

EMI. Electromagnetic interference generated by electromagnetic devices and motors—requires shielding.

Extrusion. Process used for plastics processing and metal processing. Materials are

240 METAL/POLYMER COMPOSITES

compressed in a cylinder at high temperature and forced through a small die opening which establishes the profile—which is maintained by prompt cooling.

Green strength. Term associated with compacted metal powders before they are sintered.

Hot isostatic pressing (HIP). Involves high temperature pressing of metal powders, under controlled pressures, at sintering temperatures.

Hydrogen loss. Loss in weight of metal powder or of a compact caused by heating a representative sample for a specified time and temperature in a purified hydrogen atmosphere—broadly, a measure of oxygen content of the sample (not applicable to hydride-forming elements).

Ion implantation. Ions of elements, including metals, are directed towards specific areas on substrates.

Magnetic tapes. Fine magnetic particles dispersed on polymer film.

Magnetite. Oxide of iron—Fe_3O_4—with residual magnetic properties.

MIM. Metal Injection Molding.

MPIA. Metal Powders Industries Association (Princeton, N.J.)

Porosity. The amount of pores (voids) expressed as a percentage of the total volume of the powder metallurgy part.

Prealloyed powder. Powder consisting of two or more elements which are alloyed in the powder manufacturing process and in which particles are of the same nominal composition throughout.

Rapid solidification process (RSP). Molten metal alloys are rapidly cooled at temperature gradients of 10^5 °K per second or more.

REP. Rotating Electrode Process for preparing metal powders.

RFI. Radio Frequency Interference.

Sintering. The bonding of contiguous surfaces of metal particles in a mass of powder, such as a compact, by heating at high temperatures but at a temperature lower than the melting point of the metal particles, in a controlled atmosphere.

Thin film technology. Thin metal films (generally less than one micrometer) used in semiconductors and photovoltaic devices.

Vacuum hot pressing (VHP). Utilizes vacuum pressures through flexible membranes on hot metal powders for compaction.

Vacuum metallization. Technique for deposition of heated metallic elements which, under high vacuum, deposit on plastics or ceramic substrates.

TERMS RELATING TO PLASTICS/POLYMERS

The preparation of finely divided polymers/plastics entails expressions which are pertinent to the subject matter of this text (adapted in part from ASTM-D-883-78).

Addition polymerization. Polymerization in which monomers are joined together without the splitting off of water or other simple molecules.

Antistatic compounds. Compounds intermediate between insulators (high resistivity) and conductive compounds (low resistivity).

Aramid. More usually as polyaramid—a high strength synthetic fiber (trade name "Kevlar") used for reinforcement.

Attenuation. Reduction of force or intensity as in the decrease of an electrical signal.

Atomic dimensions. Usually of the order of several angstroms (10^{-10} meters).

Bag molding. A method of molding or laminating which involves the application of fluid pressure, usually by means of air, steam, water, or vacuum to a flexible material which transmits the pressure to the material being molded or bonded. For high temperature polymer matrices, silicone rubber serves as a flexible membrane.

Block copolymer. An essentially linear copolymer in which there are repeated sequences of polymer segments of different chemical structure—some of which may be crystalline in nature, others of which may be amorphous.

Blocking. An unintentional adhesions between plastic films or between a film and another surface usually corrected by antistatic additives.

Carbon/graphite fibers. Usually prepared from mesophase pitch or polyacrylonitrile (PAN). Used as reinforcements in composites.

Catalysts. There are chemical catalysts and radiation activated catalysts which contribute to the initiation of polymerization processes. The chemical catalysts do not necessarily become part of the polymer chain.

Charge transfer salts. Present in electroactive polymers to facilitate electron flow between energy bands.

Composite. Combinations of two or more materials which offer superior performance to either material by itself. Particles, powders, fibers, films, semiliquids, or structural components may be involved.

Condensation polymerization. Polymerization in which monomers are linked together with the splitting off of water or other simple molecules.

Curing agents. Curing agents participate in the polymerization process. They may be latent—curable only at elevated temperatures or the curing agent may be activated at room temperature. (ca. 25°C).

DSC. (Differential scanning calorimeter). Instrumentation for measuring chemical reactions by observing exothermic or endothermic (heat rise or heat input) reactions of materials—usually over a programmed temperature cycle.

Epitaxial films. A term identified with semiconductors, involving the build-up, layer by layer, of thin films of molecular dimensions. Applicable to polymers, ceramics, and metals.

Glass transition. The reversible change in an amorphous polymer or in amorphous regions of a partially crystalline polymer from (or to) a viscous or rubbery condition to (or from) a hard and relatively brittle state. The glass transition generally occurs over a relatively narrow temperature region and is similar to the solidification of a liquid to a glassy state; it is not a phase transition. The *glass transition temperature* is the approximate midpoint of the temperature range over which glass transition takes place.

Honeycomb. Hexagonal configuration for reinforcement—prepared from paper, polymer, or metal films—bonded with adhesives.

Injection molding. A technique developed for processing thermoplastics which are heated and forced under pressure into closed molds and cooled.

Matrix. The principal binder (metal or polymer) in a composite which forms a continuum, rather than discrete particles.

MMC. (Metal matrix composites).

Molecular weight (M.W.) distribution. Polymers produced during polymerization will have a range of molecular weights. Typically a molecular weight distribution consists of low molecular weight fractions which are fluid and more soluble, and high molecular weight fractions which are solid. Polymer manufacturers strive to limit the range of M.W. distribution.

Monomer. A relatively simple organic chemical compound which can react to form a polymer, such as vinyl chloride or ethylene.

Plastic(s). A material that contains as an essential ingredient one or more organic polymeric substances of high molecular weight. It is solid in its finished state and, at some stage in its manufacture or processing into finished articles, can be shaped by flow under pressure. Though rubbers and paints are not necessarily considered plastics, the matrix is essentially synthetic or natural polymers.

Plastisols. Formed from thermoplastic polymers partially solubilized in liquid plasticizers. Used for dip coating of metal parts.

Polymer. A substance consisting of molecules characterized by the repetition (neglecting ends, branch junctions, and other minor irregularities) of one or more types of monomeric units.

Prepreg. In reinforced thermosetting plastics, the admixture of resin, reinforcements, fillers, etc. (which form composites) in a web or filamentous form, ready for molding.

PVC. Polyvinyl chloride.

PVF. Polyvinyl fluroide.

PTFE. Polytetrafluoroethylene.

Resin. A solid or pseudosolid organic material, often of high molecular weight, which exhibits a tendency to flow when subjected to stress, and which usually has a softening or melting range. Natural occurring resinous substances are numerous and present in fauna and flora residues. Before synthetic polymers became available, natural resins were used in metal composites in lieu of polymers, and as precursors to twentieth century synthetic polymers.

Siemens (abbreviated as S). Reciprocal of ohm, the unit of electrical resistivity— sometimes expressed as mho. Thus the electrical conductivity of conductors or semiconductors is expressed as S/cm.

Thermoplastic. A plastic that can be repeatedly softened by heating, and hardened by cooling. In the softened state, it can be shaped by flow into articles by molding and extrusion. Many natural resins may be described as thermoplastic.

Thermosetting plastic(s). Plastics which have been cured by heat or other means, are substantially infusible and insoluble. Prior to becoming infusible, thermosetting polymers such as phenolic and melamine formaldehyde possess thermoplastic qualities which permit processing.

INDEX

Ablative coatings, 103, 132
ABS, 97, 119, 131, 178
Acetylene, 81, 84
Acetylene Black, 58, 97
Acrylic polymers, 122, 176
Adhesion during encapsulation, 56, 110
Adhesion enhancement, 96, 121
Advanced composites, 1
Aerospace Corp., 159
Alcoa, 19, 141
Alkaline cells, 229
Alloying ingredients, 35
Alnico magnets, 189, 194, 195
Aluminum films, 128
Aluminum flakes, volume resistivity, 91
Aluminum molds, 42
Aluminum oxide whiskers, 26
Aluminum, polymer composite, 142
Aluminum powder, filler, 63
Aluminum processing, 19
Aluminum reflective layer, 124
American Chemical Society, 78, 203
American Cyanamid, 141
Amine curing agent, 63
Amorphous metal powders, 16, 70, 196
Amorphous polymers, 41
Antenna dishes, 177
Anti-fouling coatings, 116
Anti-static additives, 180
Anti-static composite films, 180
Anti-static shield, 177, 179, 227
Appliances, coatings, 117
Aqueous dispersions, 28, 34
Arc spray of metal, 118
Arco Solar, 234
Armor, 5, 6
Armor plate, 139
ASM, 13, 139

A.S.T.M., 14, 15, 18, 147, 167
Atomization, 19, 20
Attenuation of radiation, 166
Automobile tire, fiber reinforced, 149
Automotive chassis, composite, 136, 140
Automotive requirements, 65, 105, 106
Automotive trends, 151

B/H curves of magnets, 190, 194
Band gaps, 82
Barrier films to gases, 181, 182
Battacharya, S., 13
Battelle, 90
Batteries, energy storage, 229, 230, 237
Batteries, lead-acid, 122
Baughman, R.N., 82
Bell Aircraft, 137
Bigg, D., 89, 90 91, 92, 167
Bipolar semiconductor, 211, 222
Bismaleimides, 138
Bitumen pitch, 3, 10, 142, 145
Blending processes, 28, 32
Blending in solution, 33
Blown thermoplastic films, 44
Boats, fiber reinforced coating, 114
Body armor, 142
Body solders, 65, 66
Bonding steel, polyethylene, 145
Boron fibers, 14, 25, 146
Boron powders in nuclear environment, 180
Braid reinforcement, 150
Braided glass reinforcement, 107, 148
Brake discs, 69
Brass inserts, 58
Bronze alloys, 3
Brunswick Co., 26
Bubble domains, 204

243

Cable termination, 154
Cabot Corp., 98
Calcium sulfate, 61
Capacitors, aluminum, 232
Carbon black, 95, 97, 98, 99, 171
Carbon mats for shielding, 178
Carbon zinc battery, 229
Carbonyls, decomposition, 16, 20, 22
Carborundum Company, 26
Carlson, C., 51
Cast phenolic resins, 61
Cathode sputtering, 103, 127
Ceramic boards, 227
Ceramic capacitors, 232, 233
Ceramic fibers, 136
Ceramic lined cylinders, 160
Ceramics, 1, 24, 96, 104
Certrifugal atomization, 21
Charge transfer compounds, 80
Chariot wheels, 4
Chemical doping, 84
Chemical vapor deposition, 128
Chip carriers, 223, 228
Chromium oxide particles in magnetic materials, 201
Ciba Geigy, 55, 113, 224
Circuit board coating, 112
Circuit boards, molded, 225
Circuit protection, 95, 96
Cloisonne, 7, 8
Coal tar pitch, 142
Coating, solvent free, 105
Coatings as composites, 102
Coaxial cable, 183
Cobalt vapor, 131, 205
Coercive force, magnets, 190, 200, 202
Coercivity, 202
Cold molding, 41
Collagen, 2
Colloid particles, 27
Colloidal dispersions, 112, 121
Commutator, molded, 59
Compaction of metal powders, 44, 45
Comparison, metal processing with plastics, 17
Composite armor, 6
Composite structures on Boeing plane, 138
Composite tooling, 137
Composites and combinations, 160
Compton photoelectrons, 173

Computer shielding, 163, 169
Condensation reactions, 60
Conductive thermoplastics, 94, 171
Conformal coatings, 123
Copper artifacts, 2, 3, 5, 10
Copper clad composites, 145
Copper, electrodeposition, 120
Copper oxide, anti-fouling, 69
Copper oxide in polymer, 85
Copper powder, 23, 24
Copper strips, 59
Corona discharge for adhesion, 145
Coupling agents, 56, 111, 121, 144
Cowan, D., 78
Cryogenic composite tanks, 146
Crystalline segments of polymers, 41, 70
Curie temperatures for ferromagnets, 192, 193, 196
Czochralski system, 215

Daido Steel Co., 22
Decibel, definition, 167
Decomposition of polymers, 71
Degradation of polymers, 3, 58
Delmonte, J., 13, 71, 87, 164
Delsen Testing Lab., 226
Dendritic structures, 16, 41
Design recommendations for metal powders, 46, 48
Devcon Corp., 64, 65
Device shielding, 170
Diallyl phthalate polymer in magnets, 206
Die casting, zinc, 46
Dielectic coatings, 173
Diffusion pumps, 129
Digital recording, 199
Dimensions submicron, 12
Dip coating, 54
Dispersion coatings, 111
Dopants for polymers, 81, 82, 83
Dopants in semiconductors, 212
Doped polyacetylene, 80
Dow Chemical, 47, 226
Drive shaft, composite, 150
Dry blending, 32
D.S.C. tests on metals/polymers, 71, 72, 73, 74, 214
du Pont Co., 79, 112, 146, 157, 220, 229
Ducting, steel and aluminum, 105

Eddy currents, 193, 197, 198
Electrical conductivity chart, 78
Electrical insulation, 53, 78
Electroactive polymers, 77, 79, 81, 83
Electroconductive composites, 77, 81, 82, 89, 93, 94
Electrodeposition of iron, 23
Electroless coatings, 111, 174, 224
Electroless copper coating, 171
Electroless metal deposition, 119, 121
Electrolytic copper, 93
Electrolytic powders, 16
Electrolytic processes, 20
Electromagnetic shields, 176
Electron beam, 130
Electron micrograph, 221
Electronic circuit boards, 145
Electronic noise pollution, 163
Electrophoretic coating, 106, 113, 122, 173
Electroplating, 103, 119
Electroplating on plastics, 119
Electrostatic discharge, 176, 178, 179, 182
Embossing metal foils, 130, 145
EMI shielding, 50, 94, 118, 121, 163, 167, 170, 173
Emulsions of polymers, 103
Encapsulant objectives, 56
Encapsulation of metal components, 54
Energy bands, 77, 79, 89, 90, 217
Engineering thermoplastics, 29
Environmental resistance, 62
Epitaxial deposits, 196, 218
Epitaxial films, 16
Epoxidized phenolics, 55
Epoxy adhesives, 115, 144
Epoxy anhydride coating, 105
Epoxy coatings, 116
Epoxy coatings on steel pipe, 114
Epoxy molded components, 57
Epoxy mortar, 64
Epoxy pastes for eroded metals, 67, 68
Epoxy resins, 54, 61, 62, 105
Ermabond process, 51
Exotherm during cure, 62
Exotherm spikes in D.S.C., 71
Explosive Compaction, 46
Extruder, twin screw, 29, 30
Extrusion mixing, 28, 31, 52
Extrusion molding, 42
Extrusion of reinforced copper, 96

Fabrics, electrical conductive, 96
Fabrics, 11
Fabrics, pre impregnated, 105
FCC regulations, 164, 168
Federal Test 101C, static discharge, 178
Ferrites, barium and strontium, 189, 202
Ferrites, nickel, zinc, 200
Ferro Corp., 117, 174
Fiber aspect ratio, 90, 92
Fiber fragmentation, 35
Fiber dispersions, 34
Fiber hybrids, 136
Fiber reinforced coatings, 104
Fiber reinforcements, 13, 14, 25, 77, 106, 135
Fiber wound structure, 147
Fibers, metal coated, 23
Fibers, nylon, 83
Filament winding, 105, 136, 146
Filigrees, 7
Films, metal and polymer, 15
Fire resistance, 132
Flaked metals, 33, 58
Flakes, aluminum, 18, 91, 168
Flexible polymers, 86
Flexible printed circuits, 122
Flexible recording disks, 200
Fluid bed coatings, 116
Foam, polyurethane, 158
Foamed product, 132
Food packaging, composite films, 180
Formability of composites, 143

Galli, E., 98
Gallium arsenide, 212, 213
Galvanized steel rod and epoxy, 64
Gamma iron oxide in magnets, 201, 202, 204
Gamma radiation shielding, 184
Gas atomization, 20
Gaskets for shielding, 170
General Motors, 231, 236
Generator, electrical, 154
Germanium, 212
Gerteisen, S., 170, 174
Glass fabrics, 105
Glass fiber tape, 58
Glass fibers, metallized, 171
Glass fibers, spray-up, 116
Glass panels, reflective, 124
Glass transition temperature, 40

Gold coatings, 90, 93, 123
Gold films, light transmission, 126
Gold leaf, 4
Goodrich, B. F., 109
Graphite fibers, 14, 104, 106
Green strength, 17, 45
Grinding techniques, 18, 27
Grubbs, R., 85

Hall effect, 213
Hausner, H., 13
Heat conductivity, 53, 50
Heeger, A., 79
Helical winding, 148
Helmets, 5
Hemp fibers, 10
High frequency communications, 165
High frequency radiation, 51, 131
High frequency radiation spectrum, 164
High modulus fibers, 25
High impact composites, 142, 143
High strength fibers, 135
High temperature resins, 14, 139, 155
High vacuum equipment, 128, 129
Historical perspectives, 2
Hitachi, 213, 228
Hoeganaes Corp., 48
Holzberg, M., 150, 153
Honeycomb construction, 157, 159
Honeycomb reinforcement, 137, 138, 140, 141
Hoses, reinforced, 149, 179
Hull smoothing, 67
Hybrids of fiber, 136
Hydrostatic extrusion, 31
Hysteresis loops, 190, 194, 196
Hysteresis loss, 167, 193, 198

IBM, 96, 212, 220, 221
Impact strength, adhesives, 66, 67
Implanted atoms, 38
Impregnant, solvent free, 154
Impregnated fabrics, 136
Impregnated tapes, 104
Impregnation of coils, 154, 155
Imprinting metal/polymer foils, 130
Induction heating, 129
Infiltration of preforms, 158
Injection molded thermoplastics, 60, 94, 97
Injection molding magnesium, 46

Injection molding, rubber, 141
Injection molding, steel, 13, 49
Inlays of metal on plastic, 144
Inserts, metal, 58
Insulators, 77, 149
Integrated circuits, 210, 222
Interactions, metals and polymers, 70, 71, 72
Intumescent coatings, 103, 132, 133
Iodine as p-dopant, 79, 85
Ion implantation, 10, 16, 36, 77, 86, 218
Ionization processes, 36, 37
Iron filled molding compound, 51
Iron powders, 206
Iron whiskers, 26
Irradiation of surfaces, 120
Isostatic pressing of metal powder, 45
Ivory, 2, 5

"Kevlar" polyaramid, 149
Kirkwood commutators, 59
Kobe Steel Co., 21
Kohswa process, 20
KRF excimer laser, 220

Lacquers, 102, 103, 125
Laminate composite, 141
Laminations of steel plates and polymer films, 115, 136, 196
Laser annealing, 37
Laser, gallium arsenide, 217
Lead acid battery, 230
Lead filled aprons, 50
Lead particles, 9, 20
Lead solders, 66
Leaf springs, 151
Leather, 5
Liao, S., 91, 124, 169
Light transition, relation to surface resistance, 169
Liquid crystal displays, 227
Liquid polymers, 60, 64
Liquid solder bath, 225
Lithium battery, 230
Lithography sources, 221
LNP Corp., 194, 228
"Lost wax" process, 58
Low temperature grinding, 27

MacDiarmid, A., 79
"Maglev" trains, 187, 188

INDEX 247

Magnesium powder, 23, 46, 47
Magnet wire enamels, 206
Magnetic cores, 197
Magnetic ferrites, 190
Magnetic field intensity, 192
Magnetic hysteresis loop, 190
Magnetic materials, 41, 53
Magnetic materials, permanent, 190
Magnetic materials, soft, 189
Magnetic permeabitily, 167
Magnetic recording, 203
Magnetic recording, frequency response, 200
Magnetic recording head, 198, 201
Magnetic recording tapes, 33, 187
Magnetic storage, 201
Magnetic tapes, 130, 168
Mastic coatings auto chassis, 105, 106
May, C., 64
McGraw Hill Publ. Co., 183, 198
Mechanical attrition, 20
Meindl, J., 217
Melting points of alloys, 51, 120
Memory circuits, 222
Metal/ceramic composites, 159
Metal coated plastics, 102, 117, 131
Metal fibers, 94
Metal filled molding compounds, 41
Metal injection molding, 52
Metal inlays, 7
Metal matrices, 160
Metal matrix composites, 25
Metal powders, 7, 9, 12, 16, 17, 89
Metal powders and carbon black, 95
Metal vaporization, 128, 129
Metallized foils, 127, 131
Metallized glass panels, 168
Metallized paper, 182
Metallized thermoplastic foils, 42
Metallo-organic compounds, 2, 41, 112
Metathesis polymerization, 85
Methyl methacrylate, 60
Mica flakes, 176
Microchip production, 213
Micro-cracking, 56, 57
Microelectronic components, protection, 182
Microelectronic devices, 91
Microelectronics, 229
Microlithography, 220
Microsectioning of PCBs for analysis, 225
Microwave radiation, 166

Microwave radio, 183
Milewski, J., 26, 90
Milling rolls, 28, 32
Miniaturization of electronic devices, 216
Mirror-like metal finish, 123
Mitsubishi, 69
Mixers, high intensity, 33
Mixers, low intensity, 32, 33
Module packaging, 224
Molded plastic cases, 164
Molecular aggregates, 2
Monomers, 27
Morris, E., 147
Morrison molded fiberglass, 150
MOS devices, 222
Motorola, 213
MPIF, 15, 44, 45, 52
Multicoaxial transmission lines, 83
Myers Engineering, 34

Nano-electronic applications, 210, 221
NASA specs on ablation, 132
Natural resin coatings, 102
NEC (Japan), 213
NEMA wire spec, 207
Neodymium-iron-boron alloy, 196
Nicalon, 160
Nickel cadmium battery, 230
Nickel plated graphite, 24, 94, 177, 195
Nickel plating, 120
Nickel zinc ferrites, 200
Niobium-tin alloys, 96
Noise abatement, 136, 140
Novak, B., 85
Nuclear radiation, 53
Nuclear reaction shielding, 184
Nuclear regulatory commission, 180
Nylon coated polyethylene, 207
Nylon 6 magnets, 194

Oersted, 192
Optical circuitry, 165
Optical data storage, 203, 204
Optical devices, coating, 124
Optical recording, 203
Organic binders, 49
Organic complexes, 112
Organosols, 108
Oriented fibers, 159
Owens-Corning fiberglass, 107

Packaging for electronics, 223, 224
Packaging materials, 123
Packaging, plastics, 57
Palladium, 123
PAN (polyacrylonitrile) fibers, 174
Paper processing, 35
Papyrus, 3
Parylene coating, 123
Pastes, metal-polymer, 65, 66, 67, 69, 94, 109, 167
Patent - 1921, metal powders, 9
Pellets, 13, 24, 41, 42
Percolation threshold, 89
Permalloy, 192
Permanent magnets, 190, 202
Permeability of magnets, 192, 197
Perpendicular magnetic recording, 199
PET-polyethylene terephthalate, 94, 115, 130, 143
Phenolic resins, high temperature, 4, 54
Photochromic liquids, 203
Photodielectric analysis, 87
Photolithography, 80
Photoresist films, 215, 219, 220
Photovoltaic devices, 210
Planetary mixers, 35
Plasma process, 21
Plasma spray, 145
Plasma spray coatings, 118
Plastic coated metals, 102
Plastic magnets, 206
Plasticizers, 108
Plastics packaging for foods, 181
Plastisols, 34, 102, 108, 109
Polarized light, 87
Polyacetylene, 80, 81, 84
Polyamide-imides, 60, 67, 157
Polyaramid honeycomb, 237
Polyaramids, 104, 105, 135, 138
Polyaromatic cyanate esters, 226
Polycarbonate, 91, 97
Polyester resins, 61, 62, 67, 105, 115, 130, 198
Polyetherether ketone (PEEK), 157
Polyethylene fibers, 142
Polyethylene films (HDPE), 181
Polyethylene films (LDPE), 181
Polyethylene, high density, 95, 104, 130
Polyethylene insulators, 183
Polyethylene powders, 27

Polyethylene, spin bonded, 182
Polyimides, 55, 96, 120, 138, 157, 199, 220, 235
Polymer electroplating, 121
Polymer films, vacuum metallization, 126
Polymer powders, 26
Polymerization, 27, 51, 84
Polymerization of acetylene, 84
Polyphenylene sulfides, 55, 86
Polypropylene, 50, 90, 104, 180, 205
Polypyrroles, 82
Polysulfides, 67
Polysulfones, 112, 122, 124
Polythiophene, 84
Polyurethane resins, 61, 62, 67, 117, 140, 202
Polyvinyl fluoride (PVF), 103, 111
Porcelain balls for grinding, 35
Porous aluminum casting, 156
Porous metal, 13, 53, 156
Porous metal composites, 136, 155
Potters wheels, 4
Powder coatings, 54, 104, 117
Powder metallurgy, 13
Powders, metals, 12, 64, 65
Precipitation of powders, 20
Preforms, 41, 158
Premix Inc., 174
Premixes, 52
Prepreg fabrics, 105
Prepreggers, 136
Pressure vessels, 104
Pre-alloys, 13
Printed circuit board, 86, 145, 224
Printed circuits, 80
Propeller shaft, laminated overlay, 114
Properties of magnetic materials, 191
Properties of steel as affected by processing, 18
PTC: current limiting, 95
PTFE in lubricants, 155
Pultruded hose, 150
Pultrusion of composites, 148, 149
Pure metal data, 93
Purity of materials, 17, 41
Putty, metal-polymer, 65, 66
P.V.C., 74, 75, 83, 108, 109, 207
Pyrrole, 83

Radar Systems, 169
Radiation damage, 37, 218, 235

Radiation resistance, 213
Radiation shielding, 163
Rapid solidification process, 16, 189
Rare earth magnets, 194
Reinforced coatings, 102
Reinforced glass fiber coating, 104, 105
Resins, natural, 3
Resistivities, 82, 85
RFI shielding, 163, 171
Rib profiles for ship hulls, 145
Rivers process, 50
Rocket nozzle coatings, 132
Rogers Corp., 153
Rohr Aircraft, 63
Rosin esters, 69
Rotating barrels, 25
Rotating electrode process, 21, 22
Rotational molding, 42
Rubber, natural, 6
Rubber reinforcement, 97

Samarium, cobalt alloy, 193
SAMPE, 13
Sandwich structures, 159
Satellite receivers, 213, 217
Satellites, 165
Schottky barrier, 217, 223, 229
Schreiber, P., 181
Screen sizes, 15, 16
Screw elements for mixing, 31, 32
Sealants against radiation, 167
Sealing of metal coatings, 126
Sealing strips, 193
Semiconductor materials comparison, 212
Semiconductors, 77, 86, 210
Shawinigan carbon black, 98
Sheet molding compounds, 171
Shielding, 50, 147, 163
Shielding by Al flakes, 173
Shielding effectiveness, 166, 172
Shielding effectiveness of conductive fibers, 175
Shielding systems, comparative table, 172
Shielding systems, compounds, 171
Shirakawa, H., 79
Shrink of polymers, 56, 62, 65, 67, 105, 154
Shuttle aircraft, 140, 146
Siemen, symbol, 81
Silica fillers, 62
Silicon carbide fibers, 14, 25, 146
Silicon carbide whiskers, 159

Silicon circuits, 214, 216
Silicon crystal growth, 216
Silicon seals, 124
Silicon solar cell, 235
Silicon steel laminates, 154
Silicon steel sheets, 190
Silicon wafers, 216, 217
Silicone oil, 206
Silicone pads, 138
Silicone polymers, 70, 84
Silver 5, 6, 16, 90
Silver films, 119
Silver migration, 97
Sintered compacts, 158
Sintered parts, impregnated, 69
Sintering, 17, 18, 45, 46, 49, 158
Sioshansi, P., 37
Sleevings of woven fibers, 107
SME, 13
(SN)x polymer, 79
Solar cells, 210, 234
Solders, 65, 69
Solid state devices, 165
Solution blending, 33
Solution polymerization, 27
Solvent free coatings, 105, 106
Sony Corp., 199
Space stations, 107
Space vehicles, 159
Spacecraft re-entry, 132
SPE, 13
SPI, 14
Spin coating, 220
Squareness ratio, 205
Stackpole Corp., 202
Stainless steel fibers, 24, 159, 177
Static charges, 53
Static eliminators, 24, 181
Steel composite sheets, 143
Steel powders, 13, 15, 63
Storage batteries, 228
Storage of pastes and paints, 70
Structural composites, 135
Styrene, 60
Sub-micron sizes, 17
Sunraycer, 236, 237
Superconductivity, 77, 96, 218, 229
Surface mount technology, 227
Surface resistance, 91, 92, 93, 180
Suspension polymerization, 27, 36

Takakura, M., 86
Tank sealants, 69
TCNQ, 79
TDK Corp., 202
Teflon (PTFE), 103, 111, 112, 113, 122, 207
Thermal decomposition of metallo-organics, 77, 85, 86, 97, 128
Thermal barriers, 103
Thermal blending, 28, 30, 116
Thermoplastic binders, 51
Thermoplastic coatings, 102
Thin film semiconductors, 219
Thin films, 91, 214, 218
Thiophenes, 83, 232
Thixotropic pastes, 66
Time Magazine, 188
Tin solders, 69
Tin-palladium compounds, 119
Titanium alloy prosthesis, 37
Titanium shells, 111
Todd, D., 31
Tooling, cast metal/polymer, 63
Tooling, laminated, 60, 136
Toray Research, 150, 151, 152
Toshiba, 213, 217
Toyota Motor, 158
Transfer molding, 42, 43, 55
Transformer, 153
Transistors, 213, 216
Transmet Co., 173
Tributyltin oxide, 114
Trotter, D., 233
Tungsten fibers, 25

Ultra high frequencies, 165
Ultraviolet light cure, 115
Union Carbide, 171
Uranium 238, 53
Urea formaldehyde foams, 133
Urethane coating, 53, 113, 138

Vacuum chambers, 126, 127
Vacuum deposition, 130
Vacuum equipment, 128, 131
Vacuum metallization, 77, 103, 123, 124, 131
Vacuum metallized films, 124, 166
Vacuum tubes, 214
Van Zant, P., 216
Vibration dampening, 139, 140
Video tapes, 202
Vinyl chloride process, 36
Vinyl plastisols, 109, 110
Vinyl polymers, powdered, 65
Voltage stress on transparent plastics, 87

Water soluble binders, 50
Weather resistance, 53
Weight savings, 139, 140
Welding fixture, 60
Werner-Pfleiderer Co., 30
Whiskers, 14, 25, 146
Wigul, F., 78
Wilson Fiberfil Inc., 174, 175
Wire braids, 170
Wire reinforcement, 5, 11
Wire, rope, 134
Wires, metal, 153
Wood furniture, 7, 10
Wound coils, 56

Xerography, 51
Xerox Corp., 51
X-ray lithography, 220
X-ray shielding, 184
X-ray spectrographs of carbon black, 97

Zinc-coated steel, 144
Zinc particles, 20, 23
Zinc spray coatings, 118, 169

JAN 2 6 1990